T0258418

Several Complex Variables

Other *Chicago Lectures in Mathematics* titles available from the University of Chicago Press:

Simplicial Objects in Algebraic Topology, by J. Peter May (1967)

Fields and Rings, Second Edition, by Irving Kaplansky (1969, 1972)

Lie Algebras and Locally Compact Groups, by Irving Kaplansky (1971)

Torsion-Free Modules, by Eben Matlis (1973)

Stable Homotopy and Generalised Homology, by J. F. Adams (1974)

Rings with Involution, by I. N. Herstein (1976)

Theory of Unitary Group Representation, by George V. Mackey (1976)

Infinite-Dimensional Optimization and Convexity, by Ivar Ekeland and Thomas Turnbull (1983)

Commutative Semigroup Rings, by Robert Gilmer (1984)

Navier-Stokes Equations, by Peter Constantin and Ciprian Foias (1988)

Essential Results of Functional Analysis, by Robert J. Zimmer (1990)

Fuchsian Groups, by Svetlana Katok (1992)

Unstable Modules over the Steenrod Algebra and Sullivan's Fixed Point Set Conjecture, by Lionel Schwartz (1994)

Topological Classification of Stratified Spaces, by Shmuel Weinberger (1994)

Chicago Lectures in Mathematics

several complex variables

Raghavan Narasimhan

The University of Chicago Press
Chicago and London

International Standard Book Number: Paper, 0-226-56817-2
Library of Congress Catalog Card Number: 75-166949

The University of Chicago Press, Chicago 60637
The University of Chicago Press, Ltd., London

Published 1971
Printed in the United States of America
99 98 97 96 95 2 3 4 5 6

This book is printed on acid-free paper.

CONTENTS

PREFACE

There are at least three parts of the theory of functions of several complex variables which are of importance in various branches of m mathematics. These are:

(1) the elementary theory, comprising Hartogs' theory, domains of holomorphy, and automorphisms of bounded domains;

(2) the local and global study of analytic sets and complex spaces; and

(3) global ideal theory, Stein manifolds, coherent analytic sheaves.

Of these the third has been dealt with in the books of Hörmander and of Gunning and Rossi and in several sets of lecture notes: Cartan [10], Malgrange [23], J. Frenkel [15], and others. The second has been dealt with in books by M. Hervé [20], Gunning and Rossi [16], and in the lecture notes of Cartan [11], [12], and of Narasimhan [24].

While parts of the elementary theory are treated in most books or lecture notes (Hörmander [21], Gunning and Rossi [16], Cartan [10], Malgrange [23]) I have felt that a reasonably complete account has been lacking, and I hope that the present notes will provide access to some of the most important parts of this aspect of the theory.

The material in these notes was presented in lectures during the autumn quarter of 1969 at the University of Chicago. Much of it I covered in lectures in 1967 and 1968 at the University of Geneva. The notes taken by Pierre Siegfried at Geneva have made the task of preparing the present notes much easier.

The only prerequisites are the following: elements of the theory of functions of one complex variable and of measure theory; point set topology; the implicit function and rank theorems; and, in the last chapter, the local existence and uniqueness theorems in the theory of ordinary differential equations.

1

ELEMENTARY PROPERTIES OF FUNCTIONS OF SEVERAL COMPLEX VARIABLES

Notation. We shall denote by $\mathbb{C}, \mathbb{R}, \mathbb{Z}, \mathbb{N}$, respectively, the field of complex numbers, the field of real numbers, the ring of integers, and the set of all non-negative integers. $\mathbb{C}^n, \mathbb{R}^n$ will stand, respectively, for the vector space over $\mathbb{C}$ and $\mathbb{R}$, of n-tuples $(z_1, \ldots, z_n)$, $z_j \in \mathbb{C}$, respectively $(x_1, \ldots, x_n)$, $x_j \in \mathbb{R}$. We shall look upon these as being provided with their natural topology. We set if $z = (z_1, \ldots, z_n) \in \mathbb{C}^n$,

$$\|z\| = \sqrt{(|z_1|^2 + \ldots + |z_n|^2)},$$

$$|z| = \max_{j=1,\ldots,n} |z_j|,$$

with similar notation in the case of elements of $\mathbb{R}^n$.

In general, $\alpha, \beta, \ldots$ denote n-tuples of elements of $\mathbb{N}$; thus $\alpha = (\alpha_1, \ldots, \alpha_n)$, $\alpha_j \in \mathbb{N}$. We then set

$$|\alpha| = \alpha_1 + \ldots + \alpha_n, \quad \alpha! = \alpha_1! \ldots \alpha_n! .$$

We write $\alpha \leq \beta$ if $\alpha_j \leq \beta_j$, $j = 1, \ldots, n$ and $\alpha < \beta$ if $\alpha \leq \beta$, $\alpha \neq \beta$.

If $a \in \mathbb{C}^n$, and $\rho = (\rho_1, \ldots, \rho_n)$, the polydisc with center a and (poly) radius ρ is the set

$$P(a, \rho) = \{z \in \mathbb{C}^n \mid |z_1 - a_1| < \rho_1, \ldots, |z_n - a_n| < \rho_n\}.$$

$\overline{P}(a, \rho)$ will stand for the closure of $P(a, \rho)$ in $\mathbb{C}^n$. Finally, if $z \in \mathbb{C}^n$,

$\alpha \in \mathbb{N}^n$, we set

$$z^\alpha = z_1^{\alpha_1} \cdots z_n^{\alpha_n} .$$

We shall sometimes identify $\mathbb{C}$ with $\mathbb{R}^2$ and $\mathbb{C}^n$ with $\mathbb{R}^{2n}$. We then write

$$z = x + iy \, , \; z_j = x_j + iy_j \, , x, y \in \mathbb{R}^n, x_j, y_j \in \mathbb{R}, \; i = \sqrt{-1} \, , \; j = 1, \dots, n.$$

For a continuously differentiable complex valued function g on an open set $\Omega \subset \mathbb{C}^n$, we set

$$\frac{\partial g}{\partial z_j} = \frac{1}{2}\left\{ \frac{\partial g}{\partial x_j} - i \frac{\partial g}{\partial y_j} \right\} , \quad \frac{\partial g}{\partial \bar{z}_j} = \frac{1}{2}\left\{ \frac{\partial g}{\partial x_j} + i \frac{\partial g}{\partial y_j} \right\} ,$$

$$(D^\alpha g)(z) = \frac{\partial^{|\alpha|}}{\partial z_1^{\alpha_1} \cdots \partial z_n^{\alpha_n}} g(z) .$$

If A and B are subsets of a (Hausdorff) topological space X, we write

$$A \subset\subset B$$

if A is relatively compact in B. As usual, $\overset{\circ}{A}$ is the interior, $\overline{A}$ the closure of A, $\partial A = \overline{A} - \overset{\circ}{A}$ the boundary of A.

The class of infinitely differentiable functions on an open set $\Omega \subset \mathbb{R}^n$ is denoted by $C^\infty(\Omega)$. We say also that an element of $C^\infty(\Omega)$ is C^∞.

<u>Definition 1.</u> Let Ω be an open set in $\mathbb{C}^n$ and f a complex valued function defined on Ω. We say that f is holomorphic on Ω if to every point $a \in \Omega$ there corresponds a neighborhood U and a power series

$$\sum_{\alpha \in \mathbb{N}^n} c_\alpha (z-a)^\alpha \equiv \sum_{\alpha_1, \dots, \alpha_n \geq 0} c_{\alpha_1 \dots \alpha_n} (z_1 - a_1)^{\alpha_1} \cdots (z_n - a_n)^{\alpha_n}$$

which converges to $f(z)$ for $z \in U$. We shall denote the set of functions holomorphic on Ω by $\mathcal{H}(\Omega)$.

Lemma 1 (Abel). Suppose that $\{c_\alpha\}_{\alpha \in \mathbb{N}^n}$ is a set of complex numbers such that, for some $\rho_1, \ldots, \rho_n > 0$, there is $M > 0$ for which

$$|c_\alpha| \rho_1^{\alpha_1} \ldots \rho_n^{\alpha_n} \leq M \qquad \forall \alpha \in \mathbb{N}^n .$$

Then the series

$$\sum_{\alpha \in \mathbb{N}^n} c_\alpha (z-a)^\alpha$$

converges uniformly for $|z_j - a_j| \leq \theta \rho_j$, $0 \leq \theta < 1$. Moreover, the series of derivatives

$$\sum_{\alpha \in \mathbb{N}^n} c_\alpha D^\beta (z-z)^\alpha , \quad \beta \in \mathbb{N}^n ,$$

converges uniformly for $|z_j - a_j| \leq \theta \rho_j$.

Corollary. A holomorphic function is infinitely differentiable. Moreover, if $f(z) = \sum c_\alpha (z-a)^\alpha$, we have $c_\alpha = \frac{1}{\alpha!} (D^\alpha f)(a)$.

Lemma 2. If f is holomorphic on $\Omega \subset \mathbb{C}^n$, we have $\frac{\partial f}{\partial \bar{z}_j} = 0$, $j = 1, \ldots, n$.

The converse of this theorem is true. We shall prove it in chapter 3.

Proposition 1 (the principle of analytic continuation). Let $\Omega \subset \mathbb{C}^n$ be an open connected set, and let f be holomorphic on Ω. If f vanishes on a non-empty open subset of Ω, then $f \equiv 0$ on Ω.

Proof. Let $E = \{z \in \Omega \mid D^\alpha f(z) = 0, \ \forall \alpha \in \mathbb{N}^n\}$. If we set $E_\alpha = \{z \in \Omega \mid D^\alpha f(z) = 0\}$, then E_α is closed since $D^\alpha f$ is continuous, so that $E = \bigcap_\alpha E_\alpha$ is closed.

Let $a \in E$. Then there is an open neighborhood U of a and a power series $\sum c_\alpha (z-a)^\alpha$ which converges to $f(z)$ for $z \in U$. By the corollary to Lemma 1, $c_\alpha = \frac{1}{\alpha!} D^\alpha f(a) = 0$ since $a \in E$. Hence $f(z) = 0$ for $z \in U$, hence $D^\alpha f(z) = 0$ for $z \in U$ for any $\alpha \in \mathbb{N}^n$. Thus $U \subset E$, and it follows that E is open.

Since Ω is connected, and $E \neq \emptyset$, $E = \Omega$.

Remark. If there exists $a \in \Omega$ with $D^\alpha f(a) = 0$ for all $\alpha \in \mathbb{N}^n$, then $f \equiv 0$.

Definition 2. Let Ω be an open set in $\mathbb{R}^n$ and f a (real or complex valued) function defined on Ω. We say that f is real analytic if to every $a \in \Omega$ corresponds a neighborhood U of a and a power series $\sum c_\alpha (x-a)^\alpha$ which converges to $f(x)$ for $x \in U$.

Remarks. (1) As before, a real analytic function is C^∞.

(2) The principle of analytic continuation is valid for real analytic functions: the proof is the same.

The next two propositions are immediate consequences of the Cauchy formula for functions of one complex variable.

Proposition 2 (Cauchy's formula). Let Ω be an open set in $\mathbb{C}^n$, f a function holomorphic on Ω, $a \in \Omega$ and let $\rho = (\rho_1, \ldots, \rho_n)$, $\rho_j > 0$ be such that $\overline{P}(a,\rho) \subset \Omega$. Then, for $z \in P(a,\rho)$, we have

$$f(z) = (2\pi i)^{-n} \int_{|\zeta_1 - a_1| = \rho_n} \cdots \int_{|\zeta_n - a_n| = \rho_n} f(\zeta_1, \ldots, \zeta_n) \prod_{j=1}^{n} (\zeta_j - z_j)^{-1} d\zeta_1 \cdots d\zeta_n .$$

Corollary 1. Under the above hypotheses,

$$D^{\alpha}f(z) = \alpha!\,(2\pi i)^{-n} \int_{|\zeta_j - a_j| = \rho_j} f(\zeta)\prod_{j=1}^{n} (\zeta_j - z_j)^{-\alpha_j - 1}\, d\zeta_1 \cdots d\zeta_n .$$

We have used here a notation we shall use often later, viz.,

$$\int_{|\zeta_j - a_j| = \rho_j} \quad \text{for} \quad \int_{|\zeta_1 - a_1| = \rho_1} \cdots \int_{|\zeta_n - a_n| = \rho_n} .$$

Corollary 2. If f is holomorphic in the polydisc $P(a, \rho)$, the series

$$\sum \frac{1}{\alpha!} D^{\alpha} f(a)(z-a)^{\alpha}$$

converges to $f(z)$ for $z \in P(a, \rho)$.

Note. If φ is continuous on the set $|\zeta_j - a_j| = \rho_j$, $j = 1, \ldots, n$, the function $f(z) = \int_{|\zeta_j - a_j| = \rho_j} \varphi(\zeta) \prod (\zeta_j - z_j)^{-1} d\zeta_1 \cdots d\zeta_n$ is holomorphic in $P(a, \rho)$ and equals

$$\sum c_{\alpha} (z-a)^{\alpha} ,$$

where

$$c_{\alpha} = \int_{|\zeta_j - a_j| = \rho_j} \varphi(\zeta) \prod_j (\zeta_j - z_j)^{-\alpha_j - 1}\, d\zeta_1 \cdots d\zeta_n .$$

This follows at once from the formula

$$\prod_{j=1}^{n} (\zeta_j - z_j)^{-1} = \sum_{\alpha \in \mathbb{N}^n} \prod_{j=1}^{n} \frac{(z_j - a_j)^{\alpha_j}}{(\zeta_j - a_j)^{\alpha_j + 1}} ;$$

the series converges uniformly for z in a compact subset of $P(a, \rho)$ and $|\zeta_j - a_j| = \rho_j$.

Proposition 3 (Cauchy's inequalities). If f is holomorphic on Ω and $\overline{P}(a, \rho) \subset \Omega$, we have

$$|D^\alpha f(a)| \leq M \cdot \alpha! \, \rho^{-\alpha} \quad \text{where} \quad M = \sup_{|\zeta_j - a_j| = \rho_j} |f(\zeta)| .$$

This follows at once from Proposition 2 if we write $\zeta_j - a_j = \rho_j e^{i\theta_j}$.

Proposition 4 (the open mapping theorem). Let Ω be a connected open set in $\mathbf{C}^n$ and let f be a holomorphic function on Ω. Suppose that f is not constant. Then the map $f: \Omega \to \mathbf{C}$ is open, i.e., f maps open sets in Ω onto open sets in $\mathbf{C}$.

Proof. (a) Case $n = 1$. We may suppose that $0 \in \Omega$, and that $f(0) = 0$. It is enough to prove that $f(\Omega)$ is a neighborhood of 0. Let $\rho > 0$ be so chosen that $\{z \in \mathbf{C} \mid |z| \leq \rho\} \subset \Omega$ and $f(z) \neq 0$ for $|z| = \rho$. Let $\delta = \inf_{|z|=\rho} |f(z)|$. Then $\delta > 0$. Let $w \in \mathbf{C}$, $w \notin f(\Omega)$ and $|w| < \delta$. Then $\varphi(z) = (f(z) - w)^{-1}$ is holomorphic on Ω and, by Proposition 3, we have

$$\frac{1}{|w|} = |\varphi(0)| \leq \sup_{|z|=\rho} |\varphi(z)| \leq \frac{1}{\delta - |w|} ,$$

so that $|w| \geq \frac{1}{2}\delta$. Thus the disc $\{w \in \mathbf{C} \mid |w| < \frac{1}{2}\delta\} \subset f(\Omega)$.

(b) The general case. Let $a \in \Omega$ and let U be a convex neighborhood of a, $U \subset \Omega$. By Proposition 1, $f|U \not\equiv f(a)$. Let $b \in U$, $f(b) \neq f(a)$, and consider $D = \{z \in \mathbf{C} \mid a + z(b-a) \in U\}$. Let $g(z) = f(a+z(b-a))$, $z \in D$. Then D is a convex set containing $0, 1$ and $g(0) = f(a) \neq f(b) = g(1)$. Hence, by (a), $g(D)$ is a neighborhood of $f(a)$. Since $F(U) \supset g(D)$, the result follows.

Corollary 1 (the maximum principle). Let Ω be a bounded open connected set in $\mathbf{C}^n$ and let f be holomorphic on Ω. Set

$M = \sup_{\zeta \in \partial\Omega} \overline{\lim_{z \to \zeta, z \in \Omega}} |f(\zeta)|$. Then, if f is non-constant, we have $|f(z)| < M$ for any $z \in \Omega$.

Proof. We may suppose that $M < \infty$. The function φ on $\overline{\Omega}$ defined by

$$\varphi(z) = |f(z)|, \; z \in \Omega, \quad \varphi(\zeta) = \overline{\lim_{z \to \zeta, z \in \Omega}} |f(z)|$$

is upper semi-continuous on the compact set $\overline{\Omega}$, hence bounded. Hence $f(\Omega) = U$ is a bounded open set (if f is non-constant) by Proposition 4. Moreover, since f is open, any $w \in \partial U$ is of the form $w = \lim f(z_\nu)$, where $\{z_\nu\}$ is a sequence of points converging to a point of $\partial\Omega$. Hence $\partial U \subset \{w \in \mathbb{C} \mid |w| \leq M\}$. Since U is bounded and open, $U \subset \{w \in \mathbb{C} \mid |w| < M\}$, q. e. d.

Corollary 2. If f is holomorphic on the connected open set Ω in $\mathbb{C}^n$ and if there is $a \in \Omega$ with $|f(z)| \leq |f(a)|$ for all $z \in \Omega$, then f is constant.

Proof. If f were nonconstant, $f(\Omega)$ would be an open set contained in $\{w \in \mathbb{C} \mid |w| \leq |f(a)|\}$, hence would be contained in $\{w \in \mathbb{C} \mid |w| < |f(a)|\}$, which is absurd.

Proposition 5 (Weierstrass' theorem). Let Ω be an open set in $\mathbb{C}^n$ and let $\{f_\nu\}$ be a sequence of holomorphic functions on Ω which converges uniformly on every compact subset of Ω. Then $f = \lim f_\nu$ is holomorphic on Ω and, for $\alpha \in \mathbb{N}^n$, $\{D^\alpha f_\nu\}$ converges to $D^\alpha f$, uniformly on compact subsets of Ω.

Proof. Let $a \in \Omega$, and $\rho = (\rho_1, \ldots, \rho_n)$ be so that $\overline{P}(a, \rho) \subset \Omega$, and let $|z_j - a_j| \leq \frac{1}{2} \rho_j$. Then

$$f(z) = \lim f_\nu(z) = \lim (2\pi i)^{-n} \int_{|\zeta_j - a_j| = \rho_j} f_\nu(\zeta) \prod (\zeta_j - z_j)^{-1} d\zeta_1 \cdots d\zeta_n$$

$$= (2\pi i)^{-n} \int_{|\zeta_j - a_j| = \rho_j} f(\zeta) \prod (\zeta_j - z_j)^{-1} d\zeta_1 \cdots d\zeta_n .$$

By the note after Corollary 2 to Proposition 2, f is holomorphic in a neighborhood of a. Moreover, for $|z_j - a_j| \le \frac{1}{2}\rho_j$,

$$\frac{1}{\alpha!} D^\alpha f(z) = (2\pi i)^{-n} \int_{|\zeta_j - a_j| = \rho_j} f(\zeta) \prod (\zeta_j - z_j)^{-\alpha_j - 1} d\zeta_1 \cdots d\zeta_n$$

$$= \lim (2\pi i)^{-n} \int_{|\zeta_j - a_j| = \rho_j} f_\nu(\zeta) \prod (\zeta_j - z_j)^{-\alpha_j - 1} d\zeta_1 \cdots d\zeta_n$$

$$= \lim \frac{1}{\alpha!} D^\alpha f_\nu(z) .$$

Proposition 6 (Montel's theorem). Let $\mathcal{F} = \{f\}$ be a family of holomorphic functions on Ω such that for any compact set $K \subset \Omega$, there exists $M = M_K > 0$ so that

$$|f(z)| < M \quad \text{for } z \in K, \ f \in \mathcal{F} .$$

Then any sequence $\{f_\nu\}$, $f_\nu \in \mathcal{F}$, contains a subsequence which converges uniformly on compact subsets of Ω.

Proof. For $a \in \Omega$ and $f \in \mathcal{F}$, set $\sum c_\alpha(f, a)(z-a)^\alpha = f(z)$ in a neighborhood of a. By Corollary 2 to Proposition 2, this series converges in a polydisc $P(a, \rho)$, independent of f. Moreover, by Proposition 3, and the hypothesis on $\mathcal{F}$, there is $C > 0$ such that

$$|c_\alpha(f, a)| \le C r^{-\alpha} \quad \text{if } r = (r_1, \ldots, r_n)$$

is such that $0 < r_j < \rho_j$. Hence, if $\{f_\nu\}$ is a sequence of elements of $\mathcal{F}$, we can, by the diagonal method, find a subsequence $\{f_{\nu_k}\}$ so that $c_\alpha(f_{\nu_k}, a)$ converges as $k \to \infty$ for any $\alpha \in \mathbb{N}^n$. It follows that

$\{f_{\nu_k}\}$ converges uniformly on a neighborhood of a. In fact, let $U = P(a, r')$, with $0 < r'_j < r_j < \rho_j$. Then, for $z \in U$,

$$|f_{\nu_k}(z) - f_{\nu_\ell}(z)| \leq \sum_{|\alpha| \leq N} |c_\alpha(f_{\nu_k} - f_{\nu_\ell}, a)| + 2C \sum_{|\alpha| > N} r'^{\alpha} r^{-\alpha} .$$

Since $r'_j < r_j$, this last term $\to 0$ as $N \to \infty$ (uniformly with respect to k, ℓ). Further $c_\alpha(f_{\nu_k} - f_{\nu_\ell}, a) \to 0$ for each α as $k, \ell \to \infty$. It follows that

$$\sup_{z \in U} |f_{\nu_k}(z) - f_{\nu_\ell}(z)| \to 0 \text{ as } k, \ell \to \infty .$$

If $\{U_p\}_{p = 1, 2, \ldots}$ is a sequence of open sets covering Ω such that, for each p, any sequence of elements of $\mathfrak{F}$ has a subsequence converging uniformly on U_p, it follows, again by the diagonal method, that any sequence $\{f_\nu\}$, $f_\nu \in \mathfrak{F}$ has a subsequence $\{f_{\nu_k}\}$ converging uniformly on U_p for each p. This sequence obviously converges uniformly on any compact subset of Ω.

Definition 3. A subset A of a connected open set $\Omega \subset \mathbb{C}^n$ is called a set of uniqueness if any holomorphic function f on Ω vanishing on A is $\equiv 0$.

Example. A set with $\overset{\circ}{A} \neq \emptyset$ is a set of uniqueness.

Proposition 7 (Vitali). Let $\{f_\nu\}$ be a sequence of holomorphic functions on the connected open set Ω and let A be a set of uniqueness in Ω. Suppose that $\{f_\nu\}$ is uniformly bounded on Ω ($|f_\nu(z)| < M$ for all $z \in \Omega$, all ν), and that $\{f_\nu(a)\}$ converges for any $a \in A$. Then $\{f_\nu\}$ converges uniformly on compact subsets of Ω.

Proof. If $\{f_\nu\}$ does not converge uniformly on compact subsets of Ω, we can find $K \subset \Omega$ compact, and subsequences $\{\nu_k\}$, $\{\mu_k\}$ of $\{\nu\}$ and $\delta > 0$ and $\{z_k\} \subset K$ so that

$$|f_{\nu_k}(z_k) - f_{\mu_k}(z_k)| \geq \delta .$$

By Proposition 6, replacing $\{\nu_k\}$ and $\{\mu_k\}$ by subsequences, if necessary, we can suppose that $\{f_{\nu_k}\}$ and $\{f_{\mu_k}\}$ converge uniformly on compact subsets of Ω (to holomorphic functions f and g respectively) and that $z_k \to x_o \in K$. Then

$$|f(z_o) - g(z_o)| \geq \delta > 0.$$

On the other hand, for $a \in A$,

$$f(a) - g(a) = \lim \{f_{\nu_k}(a) - f_{\mu_k}(a)\} = 0 \text{ since } \{f_\nu(a)\} \text{ converges.}$$

Since A is a set of uniqueness, $f - g \equiv 0$, a contradiction.

Proposition 8. Let B be the half-strip $a < \operatorname{Re} z < b$, $\operatorname{Im} z > 0$, in the plane $\mathbb{C}$, and let Ω' be a connected open set in $\mathbb{C}^{n-1}$. Let $\Omega = B \times \Omega'$, and let f be a bounded holomorphic function on Ω. Suppose that for some c, $a < c < b$,

$$\lim_{y \to \infty} f(c + iy, z') = g(z')$$

exists, uniformly for z' in any compact subset of Ω'. Then $f(x + iy, z') \to g(z')$ as $y \to \infty$ uniformly in the interval $a+\varepsilon \leq x \leq b-\varepsilon$ for any $\varepsilon > 0$, and z' in a compact subset of Ω'.

Proof. Let f_ν be the function holomorphic on Ω defined by

$$f_\nu(z, z') = f(z+i\nu, z') .$$

Then $\{f_\nu\}$ is uniformly bounded, and the assumption that

$$\lim_{y \to \infty} f(c + iy, z') = g(z')$$

implies that

$$\lim_{\nu \to \infty} f_\nu(z, z') = g(z')$$

on the set $A = \{(z, z') \in \Omega \mid \operatorname{Re} z = c,\ 0 < \operatorname{Im} z < 1\}$. Since A is a set of uniqueness, the result follows from Proposition 7.

ANALYTIC CONTINUATION: ELEMENTARY THEORY

It is trivial that if Ω is a connected open set in $\mathbb{C}$ and $a \in \mathbb{C} - \Omega$, there is a holomorphic function f in Ω which cannot be continued analytically to the point a ($f(z) = (z-a)^{-1}$). This is no longer true in $\mathbb{C}^n$, $n > 1$.

Theorem 1 (Hartogs). Let $P = \{z \in \mathbb{C}^n \mid |z_j| < 1\}$, $n > 1$. Let V be a neighborhood of ∂P such that $V \cap P$ is connected (∂P has a fundamental system of such neighborhoods). Then, for any holomorphic function f on V, there is a holomorphic function F on $P \cup V$ so that $F|V = f$.

Proof. Let $\varepsilon > 0$ be such that if

$$A = \{z \in \mathbb{C}^n \mid 1-\varepsilon < |z_1| < 1, |z_j| < 1, j \geq 2\} \cup \{z \in \mathbb{C}^n \mid 1-\varepsilon < |z_2| < 1, |z_j| < 1, j \neq 2\},$$

$A \subset V$. Let $z' = (z_2, \ldots, z_n)$. If $|z'| < 1$, the function $z_1 \mapsto f(z_1, z')$ is holomorphic in the ring $\{z_1 \in \mathbb{C} \mid 1-\varepsilon < |z_1| < 1\}$, so that

$$f(z_1, z') = \sum_{\nu=-\infty}^{\infty} a_\nu(z') z_1^\nu .$$

Clearly, for any $\nu \in \mathbb{Z}$, $a_\nu(z')$ is holomorphic in $P' = \{|z_2| < 1, \ldots, |z_n| < 1\}$. If now $1-\varepsilon < |z_2| < 1$, $|z_3| < 1, \ldots, |z_n| < 1$, the function $z_1 \mapsto f(z_1, z')$ is holomorphic in the disc $|z_1| < 1$, so that its Laurent series contains no negative powers of z_1; i.e., $a_\nu(z') = 0$ for $\nu < 0$ and $1-\varepsilon < |z_2| < 1$. By Chapter 1, Proposition 1, $a_\nu(z') \equiv 0$ for $\nu < 0$, $z' \in P'$. We define

$$F(z) = \begin{cases} f(z) & , \ z \in V \\ \sum_{\nu=0}^{\infty} a_\nu(z') z_1 & , \ z \in P \end{cases}$$

The latter series converges uniformly on compact subsets of P and so is holomorphic there; further, it coincides with f on a nonempty open subset of $V \cap P$, and so on the whole of $V \cap P$, since this set is connected.

Let Ω be a domain in $\mathbb{C}^n$ (open connected set). We say that Ω is a Reinhardt domain if whenever $z = (z_1, \ldots, z_n) \in \Omega$ and $\theta_1, \ldots, \theta_n \in \mathbb{R}$, we have $(e^{i\theta_1} z_1, \ldots, e^{i\theta_n} z_n) \in \Omega$.

<u>Theorem 2.</u> Let Ω be a Reinhardt domain in $\mathbb{C}^n$. Then for any holomorphic function f on Ω, there is a "Laurent series"

$$\sum_{\alpha \in \mathbb{Z}^n} a_\alpha z^\alpha$$

which converges uniformly to f on compact subsets of Ω. Moreover, the a_α are uniquely determined by f.

<u>Proof.</u> We begin by proving the uniqueness. Let $w \in \Omega$ be a point with coordinates $(w_1, \ldots, w_n)$, $w_j \neq 0$. Then, since the series converges uniformly to f on compact subsets of Ω, we may set $z_j = w_j e^{i\theta_j}$, multiply by $e^{-i(\alpha_1\theta_1 + \ldots + \alpha_1\theta_n)}$ and integrate term by term. This gives, for $\alpha \in \mathbb{Z}^n$,

$$a_\alpha = w^{-\alpha}(2\pi)^{-n}\int_{-\pi}^{\pi}\cdots\int_{-\pi}^{\pi} f(w_1 e^{i\theta_1},\ldots,w_n e^{i\theta_n})\, e^{-i(\alpha_1\theta_1+\ldots+\alpha_n\theta_n)}\, d\theta_1\ldots d\theta_n \;.$$

To prove the existence of an expansion as above, we first remark that if $D = \{z \in C^n \mid r_j < |z_j| < R_j,\ -\infty < r_j < R_j,\ j = 1,\ldots,n\}$ and f is holomorphic on D, then, by iteration of the Laurent expansion for functions of one complex variable, it follows that f has an expansion in a Laurent series.

Let $w \in \Omega$. If $\varepsilon > 0$ is small enough, since Ω is a Reinhardt domain, the set

$$D(w,\varepsilon) = \{z \in C^n \mid |w_j| - \varepsilon < |z_j| < |w_j| + \varepsilon\}$$

is contained in Ω. Since this is a set of the form D above, there is a Laurent expansion

$$\sum_{\alpha \in \mathbb{Z}^n} a_\alpha(w) z^\alpha = f(z) \;, \quad z \in D(w,\varepsilon)$$

which converges to f uniformly in a neighborhood of w. Now, if $w' \in D(w,\varepsilon)$ and $\sum a_\alpha(w')z^\alpha$ is the expansion corresponding to w' in a set $D(w',\varepsilon) \subset \Omega$, then the uniqueness assertion above shows that $a_\alpha(w) = a_\alpha(w')$.

Hence the function $w \mapsto a_\alpha(w)$ is locally constant on Ω for any $\alpha \in \mathbb{Z}^n$. Since Ω is connected, $a_\alpha(w) = a_\alpha$ is independent of w. Hence there is a series

$$\sum_{\alpha \in \mathbb{Z}^n} a_\alpha z^\alpha$$

which converges uniformly to $f(z)$ in the neighborhood of any point of Ω, hence uniformly to $f(z)$ for z in any compact subset of Ω.

Corollary 1. Let Ω be a Reinhardt domain such that for each j, $1 \leq j \leq n$, there is a point $z \in \Omega$ whose j-th coordinate is 0. Then any holomorphic function f in Ω admits an expansion

$$f(z) = \sum_{\alpha \in \mathbb{N}^n} a_\alpha z^\alpha$$

which converges uniformly on compact subsets of Ω.

Corollary 2. Let Ω be a Reinhardt domain such that for each j, $1 \leq j \leq n$, there is a point $z \in \Omega$ whose j-th coordinate is 0. Then any holomorphic function f on Ω has a holomorphic extension F to the set $\tilde{\Omega} = \{(\rho_1 z_1, \ldots, \rho_n z_n) | \; 0 \leq \rho_j \leq 1, \; (z_1, \ldots, z_n) \in \Omega\}$ (i.e., there is a unique F holomorphic on $\tilde{\Omega}$ such that $F|\Omega = f$).

Corollary 3. Let f be holomorphic in

$$r < |z_1|^2 + \ldots + |z_n|^2 < R \quad \text{where } 0 < r < R.$$

Then f can be continued holomorphically to $|z_1|^2 + \ldots + |z_n|^2 < R$.

The question arises as to whether one can construct the "domain of existence" of one or more holomorphic functions. We shall carry out the construction for one function in this chapter, and take up families of functions later.

Let $a \in \mathbb{C}^n$. Consider the set of pairs (U, f), where U is an open set in $\mathbb{C}^n$, $a \in U$, and f is holomorphic on U. Two such pairs $(U, f), (V, g)$ are said to be equivalent if there exists a neighborhood W of a, $W \subset V \cap U$ such that $f|W = g|W$. An equivalence class with respect to this relation will be called a germ of holomorphic functions at a. We shall often identify a germ with a function representing it when no confusion is likely. Note that the value at a of a germ f_a at a, $f_a(a)$, and the value at a of any derivative of f_a are well-defined.

We denote by $\mathcal{O}_a$ the set of all germs of holomorphic functions at a. $\mathcal{O}_a$ is a ring.

The sheaf $\mathcal{O}$. Consider $\mathcal{O} = \bigcup_{a \in C^n} \mathcal{O}_a$. We have a map $p: \mathcal{O} \to C^n$ defined by $p(f) = a$ if $f \in \mathcal{O}_a$. We define a topology on $\mathcal{O}$ as follows. Let $f_a \in \mathcal{O}_a$, and let (U, f) be a pair defining f_a. Let $N(U, f) = \{f_b \mid b \in U\}$ where f_b is the germ at b defined by the pair (U, f). By definition, the sets $N(U, f)$, when (U, f) runs over all pairs defining f_a, form a fundamental system of neighborhoods of f_a. $\mathcal{O}$ is called the sheaf of germs of holomorphic functions on $\mathbb{C}^n$.

Lemma 1. The map $p: \mathcal{O} \to \mathbb{C}^n$ is continuous.

Proof. Let $f_a \in \mathcal{O}_a$; then $p(f_a) = a$. Let V be a neighborhood of a, and (U, f) a pair defining f_a. Then $p(N(U \cap V, f)) \subset V$.

Lemma 2. The topology defined above on $\mathcal{O}$ is Hausdorff.

Proof. Let $f_a, g_b \in \mathcal{O}$, $f_a \neq g_b$. If $p(f_a) \neq p(g_b)$, we can find disjoint open sets Ω_a, Ω_b in $\mathbb{C}^n$ containing $p(f_a), p(g_b)$ respectively. Then $p^{-1}(\Omega_a)$, $p^{-1}(\Omega_b)$ are disjoint open sets continaing f_a, g_b respectively.

If $p(f_a) = p(g_b)$, then $f_a, g_b \in \mathcal{O}_a$. Let $(U, f), (U', g)$ be pairs defining f_a, g_b, and let V be a connected neighborhood of a, with $V \subset U \cap U'$. Then $N(V, f), N(V, g)$ are disjoint. In fact, if $x \in N(V, f) \cap N(V, g)$, then f and g coincide in a neighborhood of $p(x)$, and V being connected, $f \equiv g$ on V; in particular $f_a = g_b$, contradiction.

Lemma 3. The map $p: \mathcal{O} \to \mathbb{C}^n$ is a local homeomorphism, i.e., any $x \in \mathcal{O}$ has a neighborhood N such that $p|N$ is homeomorphic onto an open set in $\mathbb{C}^n$.

Proof. If $x \in f_a$ and $N = N(U, f)$ where (U, f) defines f_a, then $p(N) = U$. The inverse of $p|N$ is given by $f \mapsto f_b$, $b \in U$, f_b being, as before, the germ at b defined by (U, f).

Before going further, we shall develop some general properties of spaces which have the property given in Lemma 3.

Definition 1. A Hausdorff topological space X is called a manifold of (real) dimension n if every point of X has an open neighborhood homeomorphic to an open set in $\mathbb{R}^n$.

Let X be a manifold of dimension n, X' a Hausdorff space. We say that a continuous map $p: X' \to X$ is a local homeomorphism if to every point $a' \in X'$, there is an open neighborhood U' such that $p(U')$ is open in X and $p|U'$ is a homeomorphism onto U. Note that X' carries naturally then the structure of a manifold.

If $p: X' \to X$ is a local homeomorphism, we say that the triple (X', p, X) is an (unramified) domain over X. We shall speak of X' as a domain over X when it is clear from the context to which p we are referring.

Let $p: X' \to X$ be a domain over X and Y a Hausdorff topological space, $f: Y \to X$ a continuous map. A lifting of f to X' is a continuous map $f': Y \to X'$ such that $p \circ f' = f$.

Note that a lifting does not always exist, and if it does, is not necessarily unique.

Lemma 4. Let Y be connected, and let $f_1, f_2 : Y \to X'$ be two liftings of f. Suppose also that, for some $y_o \in Y$, we have $f_1(y_o) = f_2(y_o)$. Then $f_1 \equiv f_2$.

Proof. Let $E = \{y \in Y \mid f_1(y) = f_2(y)\}$. Clearly, E is closed and nonempty. If $a \in Y$, let $b = f_1(a) = f_2(a)$. Let U' be a neighborhood of b such that $p|U'$ is a homeomorphism onto an open set U of X. Let V be a neighborhood of a such that $f_1(V), f_2(V) \subset U'$, $f(V) \subset U$. Then clearly

$$f_1|V = (p|U')^{-1} \circ f = f_2|V,$$

so that E is open. Since Y is connected, $E = Y$, so that $f_1 \equiv f_2$.

Definition 2. Let X be a Hausdorff topological space. An arc γ in X is a continuous map $\gamma: I \to X$, where $I = [0,1] \subset \mathbb{R}$. A loop in X is an arc γ with $\gamma(0) = \gamma(1)$. For an arc γ, $\gamma(0)$ is called the origin, $\gamma(1)$ the extremity of γ.

Theorem 3 (the monodromy theorem). Let $I = [0,1]$ and $p: X' \to X$ be a domain over X, $a \in X$ and $a' \in p^{-1}(a)$. Let $F: I \times I \to X$ be a continuous map such that $F(0,u) = a$, $u \in I$. For $u \in I$, let γ_u be the arc $t \to F(t,u)$ in X.

Suppose that for any $u \in I$, there is a lifting γ'_u of γ_u to X' so that $\gamma'_u(0) = a'$. Then the map

$$F': I \times I \to X'$$

defined by $(t,u) \mapsto \gamma'_u(t)$ is continuous.

In particular, if $F(1,u) = b$ is independent of $u \in I$, so is $F'(1,u) = b'$. [For $F'(\{1\} \times I)$ is a connected subset of the discrete set $p^{-1}(b)$.]

Proof. Let $u_o, t_o \in I$. To prove that F' is continuous at (t_o, u_o), we proceed as follows. Let $\{U'_1, \ldots, U'_p\}$ be a finite covering of $\gamma'_{u_o}(I)$ by open sets U'_j such that $p_j = p|U'_j$ is a homeomorphism onto $U_j \subset X$. Let $0 = \tau_o < \ldots < \tau_p = 1$ be points of I such that $\gamma'_{u_o}(t) \in U'_j$ for $\tau_{j-1} \leq t \leq \tau_j$. (The U'_j and the τ_j can be chosen so that this is satisfied.) Let $\varepsilon > 0$ be so that $\gamma_u(t) \in U_j$ for $\tau_{j-1} \leq t \leq \tau_j$, $|u-u_o| < \varepsilon$. We assert that $\gamma'_u(t) = p_j^{-1} \circ F(u,t)$ for $\tau_{j-1} \leq t \leq \tau_j$, $|u-u_o| < \varepsilon$, $j = 1, \ldots, p$. In fact, γ'_u, and $b \to p_1^{-1} \circ F(u,t)$ are two liftings of $t \mapsto F(u,t)$, $\tau_o \leq t \leq \tau_1$, which are equal to a' for $t = 0$, hence, by Lemma 4,

$$\gamma'_u(t) = p_1^{-1} \circ F(u,t) \quad \text{for } \tau_o \leq t \leq \tau_1 .$$

Suppose now that we have proved that for some j, $1 \leq j < p$,

$$\gamma'_u(t) = p_j^{-1} \circ F(u,t) \quad \text{for } \tau_{j-1} \leq t \leq \tau_j ,\ |u-u_o| < \varepsilon .$$

Then, clearly $\gamma'_{u_o}(\tau_j) = p_j^{-1} \circ F(u_o, \tau_j) = p_{j+1}^{-1} \circ F(u_o, \tau_j)$. Moreover, $p_j^{-1} \circ F(u, \tau_j)$, $p_{j+1}^{-1} \circ F(u, \tau_j)$ are two liftings of $u \mapsto F(u, \tau_j)$, $|u-u_o| < \varepsilon$ which coincide for $u = u_o$, hence, by Lemma 4, for $|u-u_o| < \varepsilon$. Hence $\gamma'_u(\tau_j) = p_{j+1}^{-1} \circ F(u, \tau_j)$, $|u-u_o| < \varepsilon$. It follows, from Lemma 4, that $\gamma'_u(t) = p_{j+1}^{-1} \circ F(u,t)$ for $\tau_j \leq t \leq \tau_{j+1}$, $|u-u_o| < \varepsilon$. We deduce, by induction that

$$F'(u,t) = \gamma'_u(t) = p_j^{-1} \circ F(u,t) \quad \text{for } \tau_{j-1} \leq t \leq \tau_j,\ |u-u_o| < \varepsilon ,\ j = 1, \ldots, p.$$

It follows at once that F' is continuous at (t, u_o) for any $t \in I$.

We shall see later how this theorem can be applied to complex analysis.

<u>Definition 3.</u> Let $p: X' \to X$ be a domain over X. We say that p is a covering if every point $a \in X$ has a neighborhood U such that $p^{-1}(U)$ is the disjoint union of open sets U'_j such that $p|U'_j$ is a homeomorphism onto U for any j.

<u>Definition 4.</u> We say that the manifold X is simply connected if it is connected and, for every loop $\gamma: I \to X$, $I = [0,1]$, there exists a continuous map $F: I \times I \to X$ such that

$$F(t,0) = \gamma(t), \quad F(t,1) = \gamma(1) = \gamma(0) \qquad \text{for all } t \in I;$$

$$F(0,u) = F(1,u) = \gamma(0) \qquad \text{for all } u \in I.$$

<u>Proposition 1.</u> Let $p: X' \to X$ be a covering, let $a' \in X'$ and let $a = p(a')$. Then for any arc $\gamma: I \to X$ with $\gamma(0) = a$, there is a lifting $\gamma': I \to X'$ with $\gamma'(0) = a'$.

<u>Proof.</u> We can find points $0 = t_o < t_1 < \ldots < t_p = 1$ in I and open sets $U_j \subset X$ such that

$$\gamma(t) \in U_j \quad \text{for } t_j \leq t \leq t_{j+1}, \; j = 0, \ldots, p-1;$$

$$p^{-1}(U_j) = \bigcup_i U'_{i,j} \quad \text{where the union is disjoint and}$$

$p_{ij} = p|U'_{i,j}$ is a homeomorphism onto U_j for each i.

We define $\gamma'(t)$, $t_o \leq t \leq t_1$ by the formula $\gamma'(t) = p_{i0}^{-1}\,\gamma(t)$, where i is so chosen that $a' \in U'_{i,0}$. If γ' is already defined for $0 \leq t \leq t_j$ $(j < p)$, we define $\gamma'(t)$, $t_j \leq t \leq t_{j+1}$, by the formula

$$\gamma'(t) = p_{k,j}^{-1}\,\gamma(t), \text{ where } k \text{ is so chosen that } \gamma'(t_j) \in U'_{k,j}.$$

<u>Corollary.</u> If $p: X' \to X$ is a covering, if X, X' are connected and if X is simply connected, then p is a homeomorphism.

<u>Proof.</u> <u>p is surjective.</u> Let $a \in p(X')$ and $b \in X$. Let γ be an arc with $\gamma(0) = a$, $\gamma(1) = b$ and γ' be a lifting of γ. Then $p(\gamma'(1)) = b$.

<u>p is injective.</u> Let $b', c' \in X'$, $a = p(b') = p(c')$. Let γ' be an arc with $\gamma'(0) = b'$, $\gamma'(1) = c'$. Let γ be the loop $p \circ \gamma'$ in X and $F: I \times I \to X$ a continuous map with the properties listed in Definition 4. By Theorem 3 and Proposition 1, there is a continuous map $F': I \times I \to X$ with $F'(0,0) = b'$, $p \circ F' = F$. By Lemma 4, $F'(t,0) = \gamma'(t)$. Since $F(1,u) = F(0,u) = a$ is independent of u, so are $F'(0,u)$ and $F'(1,u)$. Since $F(t,1)$ is independent of t, by Lemma 4, so is $F'(t,1)$. Hence $c' = \gamma'(1) = F'(1,0) = F'(1,u) = F'(1,1) = F'(t,1) = F'(0,1) = F'(0,u) = F'(0,0) = \gamma'(0) = b'$, so that p is injective.

<u>Proposition 2.</u> Let $p: X' \to X$ be a domain over X, and let X, X' be connected. Then p is a covering if and only if for any $a' \in X'$, $a = p(a')$ and any arc $\gamma: I \to X$ with $\gamma(0) = a$, there is a lifting $\gamma': I \to X'$ with $\gamma'(0) = a'$.

<u>Proof.</u> This follows at once from Theorem 3, Proposition 1 and the proof given above since any point $a \in X$ has a neighborhood U homeomorphic to a convex set in $\mathbb{R}^n$ so that U is simply connected.

<u>Proposition 3.</u> Let $p: X' \to X$ be a covering, Y a simply connected manifold and $f: Y \to X$ a continuous map. If $a' \in X'$ and $p(a') = a = f(y_o)$ $(y_o \in Y)$, then f has a lifting $f': Y \to X'$ with $f'(y_o) = a'$.

<u>Proof.</u> Let $Z = Y \times X'$ and $Y' = \{(y, x') \in Z \mid f(y) = p(x')\}$. Let $\pi: Y' \to Y$ be the map $(y, x') \to y$. We claim that $\pi: Y' \to Y$ is a

covering. In fact, let $y \in Y$, $f(y) = x$. Let U be a neighborhood of x in X such that $p^{-1}(U) = \bigcup U_j'$, where the U_j' are disjoint and $p_j = p|U_j'$ is a homeomorphism onto U. Let V be a neighborhood of y in Y so that $f(V) \subset U$. Then $\pi^{-1}(V) = \bigcup V_j'$, where $V_j' = \{(y, z') \in Z | x' = p_j^{-1} \circ f(y), y \in V\}$ is the graph of the continous map $p_j^{-1} \circ f$ of V into X'. Clearly $\pi'|V_j'$ is a homeomorphism onto V.

Let Y_0' be the connected component of Y' containing (y_0, a'). Then $\pi|Y_0'$ is also a covering. Since Y, Y_0' are connected, and Y is simply connected, by the corollary to Proposition 1, $\pi_0 = \pi|Y_0'$ is a homeomorphism. If π' is the restriction to Y_0' of the map $(y, x') \mapsto x'$ of Z into X', we may define

$$f'(y) = \pi' \circ \pi_0^{-1}(y) .$$

Corollary. If Ω is a simply connected domain in $\mathbb{C}^n$ and f is a function holomorphic on Ω and everywhere $\neq 0$, there is a holomorphic function g on Ω with $e^g = f$.

Proof. Let $\mathbb{C}^* = \mathbb{C} - \{0\}$. The map $p: \mathbb{C} \to \mathbb{C}^*$, $p(z) = e^z$ is a covering. f is, by assumption, a map $f: \Omega \to \mathbb{C}^*$. By the above proposition, there is a continuous map $g: \Omega \to \mathbb{C}$ with $f = p \circ g = e^g$. g is clearly holomorphic if f is.

Proposition 4 (Cauchy's theorem). Let Ω be a simply connected domain in $\mathbb{C}^n$. Then, given $g_1, \ldots, g_n$ holomorphic on Ω such that

$$\frac{\partial g_j}{\partial z_k} = \frac{\partial g_k}{\partial z_j} \quad \text{for } 1 \leq j, k \leq n, \tag{1}$$

there is a function f holomorphic on Ω such that

(2) $$\frac{\partial f}{\partial z_j} = g_j \quad , \quad j = 1,\ldots,n .$$

We need the following lemma.

Lemma 5. Let P be a polydisc, $P = \{z \in \mathbb{C}^n \,|\, |z_j - a_j| < \rho_j\}$. Then, if $g_1,\ldots,g_n$ are holomorphic on P and satisfy (1), there is f holomorphic on P such that the equations (2) hold. Moreover, f is uniquely determined up to an additive constant.

Proof. The uniqueness is obvious. To prove the existence, we proceed by induction on n. If $n = 1$, and

$$g_1(z) = \sum_{\nu=0}^{\infty} c_\nu (z-a)^\nu \qquad \text{on } P,$$

we may take $f(z) = \sum_{\nu=0}^{\infty} \frac{c_\nu}{\nu+1} (z-a)^{\nu+1}$.

Assume the lemma proved in $\mathbb{C}^{n-1}$, and let $g_n = \sum_{\nu=0}^{\infty} c_\nu(z_1,\ldots,z_{n-1}) z_n^\nu$. Let $f_n = \sum_{\nu=0}^{\infty} \frac{c_\nu(z_1,\ldots,z_{n-1})}{\nu+1} z_n^{\nu+1}$.

Then $\frac{\partial f_n}{\partial z_n} = g_n$; let $h_j = g_j - \frac{\partial f_n}{\partial z_j}$, $(j = 1,\ldots,n)$. Then $\frac{\partial h_j}{\partial z_k} = \frac{\partial h_k}{\partial z_j}$, $1 \le j, k \le n$. Since $h_n \equiv 0$, we conclude that $\frac{\partial h_j}{\partial z_n} = 0$, $j = 1,\ldots,n-1$, so that the h_j are independent of z_n. By induction hypothesis, there is f_o holomorphic on $P_o = \{z \in \mathbb{C}^{n-1} \,|\, |z_j - a_j| < \rho_j, \ j = 1,\ldots,n-1\}$ such that

$$\frac{\partial f_o}{\partial z_j} = h_j \quad , \quad j = 1,\ldots,n-1.$$

We may take $f = f_o + f_n$.

Proof of Proposition 4. Let $\mathcal{O}$ be the sheaf of germs of holomorphic functions on $\mathbb{C}^n$ and let $p: \mathcal{O} \to C^n$ be the projection. Let $\mathcal{O}^n$ be the subset of the Cartesian product $\mathcal{O} \times \ldots \times \mathcal{O}$ (n times) consisting of n-tuples $(\varphi_1, \ldots, \varphi_n)$ with $p(\varphi_1) = \ldots = p(\varphi_n)$. Then p defines a map $\pi: \mathcal{O}^n \to \mathbb{C}^n$ which is a local homeomorphism.

Let $X_o = \{(\varphi_1, \ldots, \varphi_n) \in \mathcal{O}^n \mid \frac{\partial \varphi_j}{\partial z_k} = \frac{\partial \varphi_k}{\partial z_j}, 1 \leq j, k \leq n\}$. Then X_o is open and closed in $\mathcal{O}^n$, and $\pi \mid X_o = \pi_o$ is a local homeomorphism of X_o into $\mathbb{C}^n$.

Let $\psi: \Omega \to X_o$ be the map $z \mapsto ((g_1)_z, \ldots, (g_n)_z)$, where $(g_j)_z$ denotes the germ of the function g_j at z, and let $X = \psi(\Omega)$. Then ψ is a homeomorphism of Ω onto X (its inverse is $\pi_o \mid X$), and, in particular, X is simply connected.

We have a map

$$\eta: \mathcal{O} \to X_o$$

defined by $(f)_a \mapsto (\frac{\partial f}{\partial z_1})_a, \ldots, (\frac{\partial f}{\partial z_n})_a$, where indices denote germs. We claim that η is a covering. In fact, let $(\varphi_{1,a}, \ldots, \varphi_{n,a}) \in \pi_o^{-1}(a) \subset X_o$, $a \in \mathbb{C}^n$, and P a polydisc $P(a, \rho)$ and $\varphi_1, \ldots, \varphi_n$ holomorphic functions in P defining $\varphi_{1,a}, \ldots, \varphi_{n,a}$ respectively. Let f be holomorphic on P and $\frac{\partial f}{\partial z_j} = \varphi_j$ on P. Then if $U = \{(\varphi_{1,b}, \ldots, \varphi_{n,b}) \mid b \in P\}$, each connected component V of $\eta^{-1}(U)$ is of the form $N(P, f')$, where $f' = f + c$, $c \in \mathbb{C}$. Clearly $\eta \mid V$ is a homeomorphism onto U.

Let X' be a connected component of $\eta^{-1}(X)$. Then $\eta: X' \to X$ is a covering and X', X are connected. By the corollary to Proposition 1, η is a homeomorphism.

We now define a holomorphic function f on Ω by $f(z)$ = the value at z of the germ $\eta^{-1} \circ \psi(z)$, $f'_z \in X'$, such that $p(f'_z) = z$. Clearly, $\frac{\partial f}{\partial z_j} = g_j$, $j = 1, \ldots, n$.

Theorem 4 (Poincaré-Volterra theorem). Let X be a topological space (Hausdorff) with a countable base for its open sets, and Y a connected manifold. Suppose that there is a continuous map $f: Y \to X$ such that $f^{-1}(x)$ is discrete for any $x \in X$, i.e., for any $a \in f^{-1}(x)$, there is a neighborhood U of a in Y such that $U \cap f^{-1}(x) = \{a\}$. Then Y has a countable base for its open sets.

Proof. Let $\{X_\nu\}_{\nu=1,2,\ldots}$ be a countable base of open sets in X. Consider the family $\mathfrak{U} = \{U\}$ of open sets in Y defined as follows: U is relatively compact in Y and is a connected component of $f^{-1}(X_\nu)$ for some ν. We assert that

(i) $\mathfrak{U}$ is a base of open sets in Y,

(ii) $\mathfrak{U}$ is countable.

Proof of (i). Let $a \in Y$, V an open set of Y, $a \in V$. Let W be a relatively compact open set in Y such that $a \in W \subset \overline{W} \subset V$ and $\overline{W} \cap f^{-1}f(a) = \{a\}$. Then ∂W is compact, hence so is $f(\partial W)$. Further, $f(a) \notin f(\partial W)$. Let ν be so that $f(a) \in X_\nu \subset X - f(\partial W)$, and U be the connected component of $f^{-1}(X_\nu)$ containing a. Then $U \subset W$ (since otherwise $U \cap \partial W \neq \emptyset$ and $X_\nu \cap f(\partial W) \supset f(U) \cap f(\partial W) \neq \emptyset$). In particular, U is relatively compact, which proves (i).

Proof of (ii). If $U \in \mathfrak{U}$, U is a finite union of open sets homeomorphic to open sets in $\mathbb{R}^n$. It follows that U has the following property:

(*) If $\{V_\alpha\}_{\alpha \in A}$ is a family of nonempty open subsets of U with $V_\alpha \cap V_\beta = \emptyset$ if $\alpha \neq \beta$, then A is countable.

Note that any countable union of sets in $\mathfrak{U}$ has the property (*).

Let $U_o \in \mathfrak{U}$. Define, inductively, $\mathfrak{U}_o = \{U_o\}$, $\mathfrak{U}_k = \{U \in \mathfrak{U} |$ there exists $V \in \mathfrak{U}_{k-1}$ with $U \cap V \neq \emptyset\}$.

We assert that (a) $\bigcup \mathfrak{U}_k = \mathfrak{U}$, (b) $\mathfrak{U}_k$ is countable.

(a). Let $\Omega = \bigcup U$, the union being over those $U \in \bigcup \mathfrak{U}_k$, and $\Omega' = \bigcup V$, the union being over those $V \in \mathfrak{U} - \bigcup \mathfrak{U}_k$. Then $\Omega \cup \Omega' = Y$, Ω, Ω' are open, and, by definition of the $\mathfrak{U}_k$, $\Omega \cap \Omega' = \emptyset$. Y being connected, $\Omega' = \emptyset$, so that $\bigcup \mathfrak{U}_k = \mathfrak{U}$.

(b). $\mathfrak{U}_o$ is countable. Suppose $\mathfrak{U}_{k-1}$ countable. Then let

$$\Omega = \bigcup_{U \in \mathfrak{U}_{k-1}} U .$$

For each ν, it follows from (*) that the family F_ν of connected components of $f^{-1}(X_\nu)$ which meet Ω is countable. Since $\mathfrak{U}_k \subset \bigcup_\nu F_\nu$, $\mathfrak{U}_k$ is countable.

This proves the theorem.

<u>Remark.</u> The theorem is true under weaker hypotheses on Y, e.g., that it is connected, locally compact, and locally connected. One proves, using the fact that f has discrete fibers and that X has a countable base, that each point of Y has a countable fundamental system of neighborhoods. Property (*) is then easily established, and the above proof applies.

Corollary. Any connected component of $\mathcal{O}$ has a countable base.

Definition 5. Let $p: X \to \mathbb{C}^n$ be a domain over $\mathbb{C}^n$. A continuous function $f: X \to \mathbb{C}$ is called holomorphic (relative to p) if for every $a \in X$, there is a neighborhood U such that

(i) $p|U$ is a homeomorphism onto the open set $V \subset \mathbb{C}^n$,

(ii) the function $f \circ (p|U)^{-1}$ is holomorphic on V.

We define also the derivatives $D^\alpha f$ of f by the requirement

$$D^\alpha f \circ (p|U)^{-1} = D^\alpha(f \circ (p|U)^{-1}).$$

Here U is as in (i), and the derivatives on the right are those on V.

Definition 6. Let $p: X \to \mathbb{C}^n$, $p': X' \to \mathbb{C}^m$ be domains over $\mathbb{C}^n$, $\mathbb{C}^m$ respectively. A continuous map $u: X \to X'$ is called holomorphic if, for any open set $V' \subset X'$ and f' holomorphic on V', the function $f' \circ u$ is holomorphic on $u^{-1}(V')$.

Most of the the theorems of chapter 1 extend obviously to domains over $\mathbb{C}^n$.

Let now Ω be a connected open set in $\mathbb{C}^n$ and f holomorphic on Ω, and let $p: X \to \mathbb{C}^n$ be a connected domain over $\mathbb{C}^n$. We say that f can be continued analytically to X if there is a holomorphic map $u: \Omega \to X$ and a holomorphic function g on X such that

(i) $p \circ u$ is the inclusion of Ω in $\mathbb{C}^n$;

(ii) $g \circ u = f$.

Note that g, if it exists, is unique (since g is determined by f on $u(\Omega)$ which is open in X by (i), and we can apply the principle of analytic continuation; see below).

Proposition 5 (principle of analytic continuation). Let $p: X \to \mathbb{C}^n$, $p': X' \to \mathbb{C}^m$ be domains over $\mathbb{C}^n, \mathbb{C}^m$ respectively. Let X be connected, and $f, g: X \to X'$ be two holomorphic maps. If $f = g$ on a nonempty open subset of X, then $f \equiv g$. [In particular, two holomorphic functions on X which coincide on a nonempty open subset of X are identical.]

Proof. Let E be the set of $a \in X$ such that $f = g$ in a neighborhood of a. E is clearly open. We show that it is closed. Since $f = g$ on E, we have $f = g$ on $\overline{E}$. Let $b \in \overline{E}$ and $y_0 = f(b) = g(b)$. Let V be an open neighborhood of y_0 such that $p'|V$ is a homeomorphism onto an open subset W of $\mathbb{C}^m$. Let U be an open neighborhood of b such that $p|U$ is a homeomorphism onto a polydisc P in $\mathbb{C}^n$, $U \subset f^{-1}(V) \cap g^{-1}(V)$. Then $p' \circ f \circ p^{-1}$ and $p' \circ g \circ p^{-1}$ are holomorphic maps of P into W which coincide on the open set $p(E \cap U)$ which is nonempty since $b \in \overline{E}$. Hence (by chapter 1, Proposition 1 applied to the components of these maps into $\mathbb{C}^m$) $p' \circ f \circ p^{-1} = p' \circ g \circ p^{-1}$ on U. Since $p'|V$ is a homeomorphism, it follows that $f = g$ on U. Hence $U \subset E$, hence $b \in E$ and E is closed.

Theorem 5. For any holomorphic function f on a connected open set Ω in $\mathbb{C}^n$, there is a connected domain $p: X \to \mathbb{C}^n$ and a continuation g of f to X with the following property.

For any connected domain $p': X' \to \mathbb{C}^n$ and a continuation g' of f to X', there is a holomorphic map $\varphi: X' \to X$ such that $p \circ \varphi = p'$ and $f \circ \varphi = f'$.

Proof. Let X be the connected component of $\mathcal{O}$ containing f_a where $a \in \Omega$ and f_a is the germ defined by f at a. Let $p: X \to \mathbb{C}^n$

be the restriction to X of the projection of $\mathcal{O}$ into $\mathbb{C}^n$. Let $g: X \to \mathbb{C}$ be defined as follows: $g(\varphi_z)$, where φ_z is a germ at z, equals the value at z of φ_z, and let $u: \Omega \to \mathcal{O}$ be the map $u(z) =$ germ at z of f. Clearly u is continuous; since Ω is connected, $u(\Omega) \subset X$, and we obtain a map $u: \Omega \to X$. Clearly $p \circ u =$ inclusion of Ω in C^n, $g \circ u = f$.

If $p': X' \to \mathbb{C}^n$ is another connected domain over $\mathbb{C}^n$, $u': \Omega \to X'$ and $g': X' \to \mathbb{C}$ define the continuation of f to X', we define a map $\varphi: X' \to X$ as follows.

Let $x' \in X$, $a' = p'(x')$. Let U' be a neighborhood of x' mapped homeomorphically by p' onto an open set U in $\mathbb{C}^n$, and let $h = g' \circ p'^{-1}$ on U. Let $\varphi(x')$ be the germ at a' of h. This gives us a continuous map $\varphi: X' \to \mathcal{O}$. Clearly $\varphi(u'(\Omega)) \subset X$. Since X' is connected, this gives us a map $\varphi: X' \to X$. One verifies the properties stated in the theorem very easily.

We shall see later (in chapter 6) how this construction can be generalized to families of holomorphic functions and leads to the theory of domains of holomorphy.

3

SUBHARMONIC FUNCTIONS AND HARTOGS' THEOREM

Let Ω be an open set in $\mathbb{C}$. We write $z = x + iy$, $z \in \mathbb{C}, x, y \in \mathbb{R}$. If u is a (real or complex valued) function twice continuously differentiable in Ω, we write

$$\Delta u = \frac{\partial^2 u}{\partial x^2} + \frac{\partial^2 u}{\partial y^2} = 4 \frac{\partial^2 u}{\partial z \, \partial \bar{z}} .$$

Definition 1. u is called harmonic in Ω if $\Delta u \equiv 0$.

Remark. The real and imaginary part of a harmonic function are harmonic. A holomorphic function f is harmonic (since $\Delta f = 4 \frac{\partial}{\partial z}(\frac{\partial f}{\partial \bar{z}}) = 4 \frac{\partial}{\partial z}(0) = 0$).

Definition 2. Let s be a real valued function on Ω (the value $-\infty$ is admitted, not the value $+\infty$). We say that the maximum principle holds for s if, for any $U \subset\subset \Omega$, we have

$$s(z) \leq \sup_{\zeta \in \partial U} s(\zeta) , \quad z \in U.$$

Proposition 1. The maximum principle holds for any real valued harmonic function on Ω.

This is a consequence of the following

Proposition 2. The maximum principle holds for any twice continuously differentiable real valued function u with $\Delta u \geq 0$.

Proof. It suffices to prove the result when $\Delta u > 0$ at every point of Ω. In fact, if $u_\varepsilon(z) = u(z) + \varepsilon|z|^2$, $\Delta u_\varepsilon = \Delta u + 4\varepsilon > 0$, so that

$$u(z) = \lim_{\varepsilon\to 0} u_\varepsilon(z) \le \lim_{\varepsilon \to 0}\{\sup_{\zeta\in\partial U} u_\varepsilon(\zeta)\} = \sup_{\zeta\in\partial U} u(\zeta).$$

Suppose that $\Delta u > 0$ everywhere and that $u(a) = \sup_{\zeta\in\partial U} u(\zeta)$, $a \in U$, for $U \subset\subset \Omega$; let $\sup_{z\in\overline{U}} u(z) = u(z_0)$. Then $z_0 \in U$ and the function $g(t) = u(x_0, t)$ has a maximum at $t = y_0$ $(z_0 = x_0 + iy_0)$. Hence

$$\frac{\partial^2 u}{\partial y^2}(z_0) = \frac{d^2}{dt^2} g(y_0) = \lim_{h\to 0} \frac{1}{h^2}\{g(y_0+h) + g(y_0-h) - 2g(y_0)\} \le 0.$$

In the same way $\frac{\partial^2 u}{\partial x^2}(z_0) \le 0$, so that $\Delta u(z_0) \le 0$, contradiction.

Proposition 3. Let $a \in \mathbb{C}$, $\rho > 0$ and $P(z,\theta) = P_{a,\rho}(z,\theta)$ $= \frac{1}{2\pi} \operatorname{Re}\left(\frac{\rho e^{i\theta} + (z-a)}{\rho e^{i\theta} - (z-a)}\right)$ for $|z-a| < \rho$, $0 \le \theta \le 2\pi$. Then $P(z,\theta) \ge 0$ and if $h(\theta)$ is a continuous function with $h(0) = h(2\pi)$,

$$\lim_{z\to z_0} \int_0^{2\pi} P(z,\theta)h(\theta)d\theta = h(\theta_0) \quad \text{if } z_0 = a + \rho e^{i\theta_0},$$

the limit is uniform with respect to θ_0.

Proof. We may suppose that $a = 0$, $\rho = 1$, and that $h(\theta) = \varphi(e^{i\theta})$, where φ is continuous on the circle $|z| = 1$. Now $P_{0,1}(z,\theta) = \frac{1-|z|^2}{|e^{i\theta} - z|^2}$ so that $P(z,\theta) \ge 0$. Moreover, for $|z| < 1$,

$$\int_0^{2\pi} P(z,\theta)d\theta = \operatorname{Re}\frac{1}{2\pi}\int_0^{2\pi} \frac{e^{i\theta}+z}{e^{i\theta}-z}\, d\theta = \operatorname{Re}\left\{\frac{1}{2\pi i}\int_{|\zeta|=1} \frac{\zeta+1}{\zeta-z}\frac{d\zeta}{\zeta}\right\} = 1.$$

Hence, for $|z| < 1$,

$$T = \int_0^{2\pi} P(z,\theta)h(\theta)d\theta - h(\theta_0) = \int_0^{2\pi} P(z,\theta)\{\varphi(e^{i\theta}) - \varphi(e^{i\theta_0})\}\, d\theta .$$

Let $\varepsilon > 0$. Then there exists $\delta > 0$ so that $|\varphi(e^{i\theta}) - \varphi(e^{i\theta_0})| < \varepsilon$ for

$|e^{i\theta} - e^{i\theta_0}| < \delta$. Hence, since $P \geq 0$,

$$. \quad |T| < \int_0^{2\pi} P(z,\theta)d\theta + 2M \int_{|e^{i\theta} - e^{i\theta_0}| \geq \delta} P(z,\theta)d\theta ,$$

where $M = \sup|\varphi(e^{i\theta})|$. Since $\int_0^{2\pi} P(z,\theta)d\theta = 1$ and $P(z,\theta) = \frac{1-|z|^2}{|e^{i\theta} - z|^2}$

it follows that if $|z - e^{i\theta_0}|$ tends to 0, $P(z,\theta)$ tends to 0 uniformly for $|e^{i\theta} - e^{i\theta_0}| \geq \delta$. Hence $\overline{\lim}_{z \to e^{i\theta_0}} |T| \leq \varepsilon$, and $\varepsilon > 0$ being arbitrary, the result follows.

<u>Corollary 1.</u> If u is harmonic in Ω, $a \in \Omega$ and $\{z \in \mathbb{C} \mid |z-a| \leq \rho\} \subset \Omega$, then

$$u(z) = \int_0^{2\pi} P(z,\theta)\, u(a + \rho e^{i\theta})d\theta \quad \text{for } |z-a| < \rho.$$

<u>Proof.</u> We may suppose that u is real-valued. The function $v(z) = \int_0^{2\pi} P(z,\theta)\, u(a+\rho e^{i\theta})d\theta$ is the real part of the holomorphic function $\frac{1}{2\pi}\int_0^{2\pi} \frac{\rho e^{i\theta} + (z-a)}{\rho e^{i\theta} - (z-a)}\, u(a+\rho e^{i\theta})d\theta$ in $|z-a| < \rho$, and so is harmonic. The result follows from Propositions 1 and 3.

<u>Corollary 2.</u> A harmonic function is a real analytic function of (x, y), $z = x + iy$.

<u>Definition 3.</u> Let u be a map of $\Omega \subset \mathbb{C}$ into $\mathbb{R} \cup \{-\infty\}$ so that $u(\Omega) \neq \{-\infty\}$. We say that u is upper semi-continuous (u.s.c.) if

$$\overline{\lim_{\substack{z \to a, \\ z \neq a}}}\; u(z) \leq u(a) \,; \qquad \forall \; a \in \Omega ,$$

equivalently, if the set $\{z \in \Omega \mid u(z) < \alpha\}$ is open in Ω for any $\alpha \in \mathbb{R}$.

Note that if u is u.s.c., it is bounded above on every compact $K \subset \Omega$; in fact K is contained in the union of the increasing sequence of open sets $\{z \in \Omega \mid u(z) < \nu\}$, so in one of them.

Lemma 1. Let u be u.s.c. in Ω and bounded above. Then there is a sequence $\{u_k\}$ of continuous functions on Ω such that $u_k(z)$ decreases to $u(z)$ for all $z \in \Omega$.

Proof. We set $u_k(z) = \sup_{\zeta \in \Omega} \{u(\zeta) - k|\zeta - z|\}$, $M = \sup_{\zeta \in \Omega} u(\zeta)$. Clearly $-\infty < u_k(z) < +\infty$ for all $z \in \Omega$. We have $u_k(z) \geq u(z) - k|z-z| = u(z)$. Further, since the sequence $\{u(\zeta) - k|\zeta - z|\}$ decreases as k increases (for fixed ζ, z), $u_k(z)$ decreases for every z. Further, if $z, z' \in \Omega$,

$$u_k(z) \geq u(\zeta) - k|\zeta - z| \geq u(\zeta) - k|\zeta - z'| - k|z - z'|, \quad \forall \zeta \in \Omega,$$

so that $u_k(z) \geq u_k(z') - k|z - z'|$. Hence $|u_k(z) - u_k(z')| \leq k|z - z'|$, so that u_k is continuous on Ω.

To prove that $u_k(z) \to u(z)$ as $k \to \infty$, suppose first that $u(z) > -\infty$. Let $\varepsilon > 0$ and $\Omega' = \{z' \in \Omega \mid u(z') < u(z) + \varepsilon\}$; Ω' is an open neighborhood of z, and contains a disc $|z' - z| < \delta$. Let k_o be such that $M - k_o \delta < u(z)$. Then $u(z') - k|z' - z| \leq u(z') < u(z) + \varepsilon$, for $z' \in \Omega'$, while $u(z') - k|z' - z| < M - k_o \delta < u(z)$ for $z' \notin \Omega'$, $k \geq k_o$. Hence, for $k \geq k_o$,

$$u(z) \leq u_k(z) < u(z) + \varepsilon, \quad \text{so that } u_k(z) \to u(z) \text{ as } k \to \infty.$$

If $u(z) = -\infty$, and $c > 0$, then

$$\Omega' = \{z' \in \Omega \mid u(z) < -c\}$$

contains a disc $|z' - z| < \delta$, so that

$$u_k(z) \leq \max(-c, M - k\delta)$$

as before, and $u_k(z) \to -\infty$ as $k \to \infty$.

Definition 4. Let Ω be an open set in $\mathbb{C}$ and u an u.s.c. function on Ω. Suppose that u is $\not\equiv -\infty$ on any connected component of Ω. We say that u is subharmonic if the following condition is satisfied.

For any open $U \subset\subset \Omega$ and any continuous real valued function h on $\overline{U}$ which is harmonic in U, if $u(z) \leq h(z)$ for $z \in \partial U$, then $u(z) \leq h(z)$ for $z \in U$.

Remarks. 1. $I = (a, b) \subset \mathbb{R}$, the solutions of $\Delta h = 0$ where $\Delta = \frac{d^2}{dt^2}$ are the linear functions $h(t) = \alpha t + \beta$. Functions u on I such that $u(t_o) \leq h(t_o)$, $u(t_1) \leq h(t_1)$ implies $u(t) \leq h(t)$ for $t_o \leq t \leq t_1$ (where $t_o < t_1$ belong to I and h is a linear function) are precisely the convex functions. Subharmonic functions may thus be looked upon as complex analogues of convex functions.

2. Definition 4 may be reformulated as follows.

For any open $U \subset\subset \Omega$ and any h harmonic and real valued on U, the maximum principle holds for $u - h$.

Lemma 2. Let u be subharmonic on the open set $\Omega \subset \mathbb{C}$. We have

(a) The set $\{z \in \Omega \mid u(z) = -\infty\}$ contains no non-empty open set.

(b) If $a \in \Omega$ and $\rho > 0$ is such that $\{z \in \mathbb{C} \mid |z-a| \leq \rho\}$ Ω, then

$$\int_0^{2\pi} |u(a + \rho e^{i\theta})| \, d\theta < \infty .$$

Proof. (a) Suppose the assertion false. Then, there is $a \in \Omega$ and $\rho > 0$ such that if $K = \{z \in \mathbb{C} \mid |z-a| \leq \rho\}$, we have $K \subset \Omega$, $u(z_o) > -\infty$ for some $z_o \in K$, $u(z) = -\infty$ for all z in a nonempty open subset of ∂K. Let $\{u_k\}$ be a sequence of continuous functions decreas-

decreasing to u on (a neighborhood of) K (Lemma 1). Let $h_k(z) = \int_0^{2\pi} P_{a,\rho}(z,\theta)u_k(a+\rho e^{i\theta})d\theta$. Then (Proposition 3), h_k is continuous on K and is clearly harmonic on $\overset{o}{K}$. Further, $h_k(z) = u_k(z) \geq u(z)$ if $z \in \partial K$. Hence, by definition of subharmonic functions, $u(z_o) \leq h_k(z_o)$ Since u_k decreases to u, and $P_{a,\rho}(z_o,\theta)$ is positive, we conclude that

$$-\infty < u(z_o) \leq \overline{\lim_{k\to\infty}}\ h_k(z_o) = \int_0^{2\pi} P_{a,\rho}(z_o,\theta)u(a+\rho e^{i\theta})d\theta ,$$

contradicting our assumption that $u(a+\rho e^{i\theta}) = -\infty$ for θ in a nonempty open subset of $[0, 2\pi]$ (since u is bounded above on K).

(b). Let $K = \{z \in \mathbb{C} \mid |z-a| \leq \rho\} \subset \Omega$, and let $\{u_k\}$ be a sequence of continuous functions decreasing to u on K. Let $M = \sup_{z\in K} u_1(z)$. Then $u_k(z) \leq M$ for $z \in K$. Let $u_k^- = \min(u_k, 0)$. Let $z_o \in \overset{o}{K}$ be such that $u(z_o) > -\infty$. As in (a), we have

$$\begin{aligned} u(z_o) &\leq \int_0^{2\pi} P_{a,\rho}(z_o,\theta)u_k(a+\rho e^{i\theta})d\theta \\ &\leq C\cdot M + \int_0^{2\pi} P_{a,\rho}(z_o,\theta)u_k^-(a+\rho e^{i\theta})d\theta , \end{aligned}$$

$$C = \sup_\theta P_{a,\rho}(z_o,\theta)$$

so that if $u^- = \min(u, 0)$, we have

$$\begin{aligned} \int_0^{2\pi} P_{a,\rho}(z_o,\theta)|u^-(a+\rho e^{i\theta})|\,d\theta &= \lim_{k\to\infty} -\int_0^{2\pi} P_{a,\rho}(z_o,\theta)u_k^-(a+\rho e^{i\theta})d\theta \\ &\leq C\cdot M - u(z_o) . \end{aligned}$$

Since $P_{a,\rho}(z_o,\theta) \geq \delta > 0$ for all θ, we deduce that

$$\int_0^{2\pi} |u^-(a+\rho e^{i\theta})|\,d\theta < \infty.$$

Since u is bounded above on K, the result follows.

Proposition 4. Let Ω be an open set in $\mathbb{C}$ and u an u.s.c. function on Ω such that $u \not\equiv -\infty$ on any connected component of Ω. Then u is subharmonic if and only if the following condition is satisfied:

(*) For any $a \in \Omega$, there exists $R = R(a) > 0$ such that

$$u(a) \leq \frac{1}{2\pi} \int_0^{2\pi} u(a + \rho e^{i\theta}) d\theta \quad \text{for } 0 < \rho < R(a).$$

Proof. We may suppose that Ω is connected.

(i) Suppose that u is subharmonic, and suppose that $K = \{z \in \mathbb{C} \,\big|\, |z-a| \leq R\} \subset \Omega$. Let $\{u_k\}$ be a sequence of continuous functions decreasing to u on K, and $h_k(z) = \int_0^{2\pi} P_{a,\rho}(z,\theta) u_k(a+\rho e^{i\theta}) d\theta$, $z \in \overset{\circ}{K}$, $0 < \rho < R$. Then h_k is continuous on K, harmonic on $\overset{\circ}{K}$ and $h_k(z) = u_k(z) \geq u(z)$ for $z \in \partial K$. Hence $u(z) \leq h_k(z)$, $z \in K$. Hence

$$u(z) \leq \lim_{k \to \infty} h_k(z) = \int_0^{2\pi} P_{a,\rho}(z, \theta) u(a + \rho e^{i\theta}) d\theta \quad , \quad z \in \overset{\circ}{K} ;$$

in particular, since $P_{a,\rho}(a, \theta) \equiv \frac{1}{2\pi}$, $u(a) \leq \frac{1}{2\pi} \int_0^{2\pi} u(a + \rho e^{i\theta}) d\theta$.

(ii) For the converse, because of Remark 2 after Definition 4 and Corollary 1 to Proposition 3, it suffices to prove that if u satisfies condition (*), then the maximum principle holds for u.

Let $U \subset\subset \Omega$, and suppose that $u(z_0) > \sup_{z \in \partial U} u(z)$, $z_0 \in U$. Then $\sup_{z \in \overline{U}} u(z) > \sup_{\zeta \in \partial U} u(\zeta)$. Since u is u.s.c., there exists $a \in \overline{U}$ so that $u(a) = \sup_{z \in \overline{U}} u(z)$. Clearly, $a \notin \partial U$, so that $a \in U$. We shall prove that, if (*) holds, u is constant on the connected component V of U containing a; this is impossible since $u(a) > \sup_{\zeta \in \partial U} u(\zeta) \geq \sup_{\zeta \in \partial V} u(\zeta)$.

Let $E = \{z \in V \,|\, u(z) = u(a)\}$. Then $E \neq \emptyset$, and since u is u.s.c. E is closed. It suffices to prove that E is open. Clearly $u(a) > -\infty$.

Let $b \in E$ and $R = R(b) > 0$ be so that $\{z \in \mathbb{C} \mid |z-b| \leq R\} \subset V$ and $u(a) = u(b) \leq \frac{1}{2\pi}\int_0^{2\pi} u(b+\rho e^{i\theta})d\theta$ for $0 < \rho \leq R$. If $u(b+\rho e^{i\theta_0}) \leq u(a) - \varepsilon$, $\varepsilon > 0$, then $u(b+\rho e^{i\theta}) < u(a) - \varepsilon$ for all θ in a nonempty open subset I of $[0, 2\pi]$. Hence, if μ denotes the Lebesgue measure on the real line

$$u(a) \leq \frac{1}{2\pi}\mu([0, 2\pi] - I)u(a) + \frac{1}{2\pi}\mu(I)[u(a) - \varepsilon] < u(a) ,$$

a contradiction. Hence $u(b+\rho e^{i\theta}) = u(a)$ for all $\theta \in [0, 2\pi]$. Since this holds for all ρ, $0 < \rho < R$, it follows that E is open, q.e.d.

Remark. If u is subharmonic in Ω and $\{z \in \mathbb{C} \mid |z-a| \leq \rho\} \subset \Omega$, then for $|z-a| < \rho$, we have

$$u(z) \leq \int_0^{2\pi} P_{a,\rho}(z, \theta)u(a+\rho e^{i\theta})d\theta .$$

This is part (i) of the proof of Proposition (5).

Corollary 1. If u_1, u_2 are subharmonic, $\lambda_1 u_1 + \lambda_2 u_2$ is also subharmonic for $\lambda_1 \geq 0, \lambda_2 \geq 0$.

Corollary 2. If u_1, u_2 are subharmonic, so is $u = \max(u_1, u_2)$.

Proof. If $a \in \Omega$ and $u(a) = u_j(a)$, $j = 1, 2$,

$$u(a) = u_j(a) \leq \frac{1}{2\pi}\int_0^{2\pi} u_j(a+\rho e^{i\theta})d\theta \leq \frac{1}{2\pi}\int_0^{2\pi} u(a+\rho e^{i\theta})d\theta$$

for small enough ρ.

Corollary 3. If $\{u_\alpha\}$ is a family of subharmonic functions in Ω, and if the function

$$z \longmapsto u(z) = \sup_\alpha u_\alpha(z)$$

is u.s.c. in Ω, then u is subharmonic. (Same reasoning as in Corollary 2, using the above Remark.)

Corollary 4. If u is subharmonic on Ω and $a \in \Omega$, then

$$u(a) = \overline{\lim_{z \to a, z \neq a}} u(z).$$

Proof. Let $\ell = \overline{\lim}_{z \to a, z \neq a} u(z)$. Since u is u. s. c., $\ell \leq u(a)$. If $\ell < u(a)$, we have $u(a + \rho e^{i\theta}) < \ell'$ (where $\ell < \ell' < u(a)$) for small enough $\rho \neq 0$, contradicting (*).

Corollary 5. If u is a continuous function such that u and -u are subharmonic, then u is harmonic

Proof. We see at once that $u(z) = \int_0^{2\pi} P_{a,\rho}(z, \theta) u(a + \rho e^{i\theta}) d\theta$ for ρ small enough and $|z-a| < \rho$; here $a \in \Omega$ is arbitrary.

Corollary 6 (converse of Proposition 3). If u is a continuous function such that, for all $a \in \Omega$, there exists $R(a) > 0$ such that

$$u(a) = \frac{1}{2\pi} \int_0^{2\pi} u(a + \rho e^{i\theta}) d\theta \quad \text{for } 0 < \rho < R(a),$$

then u is harmonic.

Proof. For then u and -u are subharmonic.

Corollary 7. If every point of Ω has a neighborhood in which u is subharmonic, then u is subharmonic on Ω.

Proposition 5. Let u be a twice continuously differentiable real valued function on the open set $\Omega \subset \mathbb{C}$. Then u is subharmonic if and only if $\Delta u \geq 0$ everywhere in Ω.

Proof. (i) Suppose that $\Delta u \geq 0$ on Ω. If h is harmonic on $U \subset\subset \Omega$, then

$$\Delta(u - h) = \Delta u \geq 0 \quad \text{on } U.$$

Hence, by Proposition 2, the maximum principle holds for u-h, so that u is subharmonic by Remark 2 after Definition 4.

(ii) Conversely, suppose that u is subharmonic and that $\Delta u(a) < 0$, $a \in \Omega$. Then $\Delta u(z) < 0$ for all $z \in U$, where U is an open neighborhood of a. By Part (i), $-u$ is subharmonic on U. By Corollary 5 above, u is harmonic on U, so that $\Delta u(a) = 0$, a contradiction.

Examples. Let f be a function holomorphic on the open set $\Omega \subset \mathbb{C}$.

(i) If f does not vanish identically on any connected component of Ω, then

$$u(z) = \log|f(z)|$$

is subharmonic.

Proof. We have to verify condition (*) of Proposition 5. Let $a \in \Omega$. If $f(a) = 0$, $u(a) = -\infty$, and (*) at a is trivial. If $f(a) \neq 0$, then in a disc about a in which f has no zeros, u is the real part of a holomorphic function (any g with $e^g = f$), so is harmonic, in particular subharmonic.

(ii) For any $\alpha > 0$, $u(z) = |f(z)|^\alpha$ is subharmonic on Ω.

Proof. If $f(a) = 0$, condition (*) at a is trivial. If f has no zeros in a disc D about a, then $u(z) = |g(z)|$ where $g(z) = e^{\alpha \log f(z)}$ is holomorphic in D. Hence

$$g(a) = \frac{1}{2\pi}\int_0^{2\pi} g(a+\rho e^{i\theta})d\theta, \quad \rho \text{ small;}$$

hence

$$u(a) = |g(a)| \leq \frac{1}{2\pi}\int_0^{2\pi} |g(a+\rho e^{i\theta})|\,d\theta = \frac{1}{2\pi}\int_0^{2\pi} u(a+\rho e^{i\theta})d\theta.$$

(iii) If u is continuous on Ω and, for some $a \in \Omega$, u is subharmonic on $\Omega - \{a\}$, then u is subharmonic on Ω.

Proof. For any $\varepsilon > 0$, $u_\varepsilon(z) = u(z) + \varepsilon \log|z-a|$ is subharmonic on Ω: condition (*) is obviously verified for u_ε at a, and also at any point $\neq a$ since, in a neighborhood, u and $\log|z-a|$ are subharmonic. Hence, for $z \neq a$, $|z-a| < \rho$, ρ small,

$$u(z) + \varepsilon\log|z-a| \leq \int_0^{2\pi} P_{a,\rho}(z,\theta)u_\varepsilon(a+\rho e^{i\theta})d\theta .$$

Since u and u are continuous for $z \neq a$, we get, letting $\varepsilon \to 0$,

$$u(z) \leq \int_0^{2\pi} P_{a,\rho}(z,\theta)u(a+\rho e^{i\theta})d\theta \quad \text{for } |z-a| < \rho , \ z \neq a .$$

By continuity, this holds also for $z = a$, and the result follows.

Remark. The same result, with a similar proof, applies when the single point a is replaced by any countable subset of Ω.

Proposition 6 (Hadamand's three circles theorem). Let $R > 0$ and $\Omega = \{z \in \mathbb{C} \mid 0 < |z| < R\}$. Let f be holomorphic on Ω, and, for $0 < r < R$, let $M(r) = \sup_{|z|=r} |f(z)|$. Then $\log M(r)$ is a convex function of $\log r$, i.e., $\log M(e^t)$ is a convex function of t.

Proof. Let $u(z) = \sup_{\alpha \in \mathbb{R}} \log|f(ze^{i\alpha})|$. Then, u is continuous on Ω, and $u(z) = \log M(|z|)$. By Example (i) above and Corollary 3 to Proposition 4, u is subharmonic in Ω.

Suppose now that

$$\log M(r) \leq \ell(\log r) \quad \text{for } r = r_o, r_1, \quad 0 < r_o < r_1 < R,$$

where ℓ is a linear function, $\ell(t) = \alpha t + \beta$. We have to prove that $\log M(r) \leq \ell(\log r)$ for $r_o < r < r_1$. Now

$$u(z) \leq h(z) \quad \text{for } z \in \partial U, \quad U = \{z \in \Omega \mid r_o < |z| < r_1\},$$

where $h(z) = \alpha \log|z| + \beta$ is harmonic in Ω. Since u is subharmonic, this inequality holds in U, and our result follows.

Remark. This result can be written

$$m(r) \leq M(r_o)^{\eta} M(r_1)^{1-\eta} \quad \text{for } r_o < r < r_1$$

where

$$\eta = \frac{\log r_1 - \log r}{\log r_1 - \log r_o} .$$

The following proposition is one of the main steps in a fundamental theorem of Hartogs.

Proposition 7 (Hartogs). Let $R > 0$, $D = \{z \in \mathbb{C} \mid |z| < R\}$, and let $\{u_k\}$ be a sequence of subharmonic functions in D. Suppose, in addition, that the following two conditions are satisfied:

(i) There exists $M > 0$ such that $u_k(z) \leq M$ for all $z \in D$ and all k.

(ii) $\overline{\lim_{k\to\infty}} u_k(z) \leq m$ for fixed z with $|z| = \rho$, $\rho < R$.

Then, if $r < \rho$ and $m_k = m_k(r) = \sup_{|z| \leq r} u_k(z)$, we have

$$\overline{\lim_{k\to\infty}} m_k \leq m .$$

In particular, if (ii) holds for all $z \in D$, the conclusion holds for all $r < R$.

Proof. (a) Let $\varepsilon > 0$. Then there exists $E \subset [0, 2\pi]$ with $\mu(E) < \varepsilon$ (μ denotes Lebesgue measure) and k_o such that $u_k(\rho e^{i\theta}) < m + \varepsilon$ if $\theta \notin E$, $k \geq k_o$.

In fact, let

$$E_k = \bigcup_{\ell \geq k} \{\theta \in [0, 2\pi] \mid u_\ell(\rho e^{i\theta}) \geq m + \varepsilon\}.$$

Since $\overline{\lim}\, u_k(z) \leq m$ for $|z| = \rho$, we have $\bigcap_{k=1}^{\infty} E_k = \emptyset$. Now

$E_k \subset E_{k+1}$. Hence, there exists k_o such that $\mu(E_{k_o}) < \varepsilon$; let $E = E_{k_o}$. Clearly, (a) is satisfied with E and k_o as obtained here.

(b) This is the proposition. We have, for $|z| \leq r < \rho$,

$$u_k(z) \leq \int_0^{2\pi} P_{0,\rho}(z,\theta) u_k(\rho e^{i\theta}) d\theta \ .$$

If we set $C = \sup P_{0,\rho}(z,\theta)$, $\theta \in [0, 2\pi]$, $|z| \leq r$, and $E' = [0, 2\pi] - E$ we have

$$\begin{aligned} u_k(z) &\leq C \int_E u_k(\rho e^{i\theta}) d\theta + \int_{E'} P_{0,\rho}(z,\theta) u_k(\rho e^{i\theta}) d\theta \\ &\leq MC\mu(E) + \int_{E'} P_{0,\rho}(z,\theta)(m+\varepsilon) d\theta \ , \qquad k \geq k_o \ , \\ &\leq MC\varepsilon + (m+\varepsilon) \quad \text{since} \int_{E'} P_{0,\rho}(z,\theta) d\theta \leq \int_0^{2\pi} P_{0,\rho}(z,\theta) d\theta \\ &\qquad\qquad\qquad\qquad\qquad\qquad\qquad = 1 \ . \end{aligned}$$

Thus $m_k(r) \leq MC\varepsilon + m + \varepsilon$, $k \geq k_o$, and the result follows.

Remark. If h is a function continuous on $|z| \leq \rho$ and harmonic in $|z| < \rho$, and if $\overline{\lim}_{k \to \infty} u_k(z) \leq h(z)$ for all z, $|z| = \rho$, we have, for $r < \rho$, $\overline{\lim}_{k \to \infty} \alpha_k(r) \leq 0$, where $\alpha_k(r) = \sup_{|z| \leq r} \{u_k(z) - h(z)\}$. The proof is identical; we are of course assuming (i).

We have seen in chapter 1, that if f is a holomorphic function on an open set $\Omega \subset \mathbb{C}^n$, then $\frac{\partial f}{\partial \bar{z}_j} = 0$ for $j = 1, \ldots, n$. A remarkable theorem of Hartogs asserts the converse. This can be formulated as follows.

Let Ω be an open set in $\mathbb{C}^n$ and $a_1, \ldots, a_n \in \mathbb{C}$. We denote by $\Omega_{j,a}$ the open set in $\mathbb{C}$ $\{z \in \mathbb{C} | (a_1, \ldots, a_{n-1}, z, a_{j+1}, \ldots, a_n) \in \Omega\}$. For any function f on Ω, we denote by $f_{j,a}$ the function on $\Omega_{j,a}$ defined by $f_{j,a}(z) = f(a_1, \ldots, a_{j-1}, z, a_{j+1}, \ldots, a_n)$.

Theorem of Hartogs. Let f be a function defined on Ω such that, for any $a_1, \ldots, a_{n-1} \in \mathbb{C}$ and any j, $1 \leq j \leq n$, the function $f_{j,a}$ is holomorphic on $\Omega_{j,a}$. Then f is holomorphic on Ω.

We may clearly suppose that $\Omega = \{z \in \mathbb{C}^n \mid |z_j| < R\}$, $R > 0$. The proof requires a few preliminary propositions.

Lemma 3. Suppose that Ω is the polydisc defined above and that f satisfies the conditions of Hartogs' theorem. Suppose in addition that f is bounded on Ω. Then f is holomorphic on Ω.

Proof. We remark that f is measurable on the product of circles $|\zeta_j| = \rho$. If $n = 1$, f is continuous, hence measurable. Assuming the measurability proved in $n-1$ dimensions, let $\theta_1, \ldots, \theta_p$ be points on the circle with $\theta_p = \theta_1$ and the length of the arc $\theta_j\theta_{j+1} < \varepsilon$. Then, the function

$$f_\varepsilon(\zeta_1, \ldots, \zeta_n) = f(\theta_j, \zeta_2, \ldots, \zeta_n) \quad \text{if } \zeta_1 \text{ is in the arc } \theta_j\theta_{j+1}$$

(open on the right and closed on the left) converges at each point to $f(\zeta)$ since f is continuous in ζ_1 for fixed $\zeta_2, \ldots, \zeta_n$. By induction, f_ε is measurable, hence so is f.

Suppose that $|f(z)| \leq M$. Let $0 < \rho < R$. By chapter 1, Proposition 2 applied to the case $n = 1$, we have, if $|z_j| < R$, $j = 1, \ldots, n-1$, and $|z_n| < \rho$,

$$f(z_1, \ldots, z_n) = \frac{1}{2\pi i} \int_{|\zeta_n| = \rho} \frac{f(z_1, \ldots, z_{n-1}, \zeta_n)}{\zeta_n - z_n} \, d\zeta_n .$$

Since, for fixed $z_1, \ldots, z_{n-2}, \zeta_n$, $f(z_1, \ldots, z_{n-1}, \zeta_n)$ is holomorphic in the disc, we may repeat this procedure, and we obtain, for $|z_j| < \rho$, $j = 1, \ldots, n$,

$$f(z_1,\dots,z_n) = (2\pi i)^{-n} \int_{|\zeta_n|=\rho} d\zeta_n \cdots \int_{|\zeta_1|=\rho} \frac{f(\zeta_1,\dots,\zeta_n)}{\prod(\zeta_j - z_j)}\, d\zeta_1$$

where the integral is an iterated integral. Now, for $|z_j| \le r < \rho$ and $|\zeta_j| = \rho$, we have

$$\prod(\zeta_j - z_j)^{-1} = \sum_{\alpha \in \mathbb{N}^n} \frac{z^\alpha}{\zeta_1^{\alpha_1+1} \cdots \zeta_n^{\alpha_n+1}},$$

and the series converges uniformly for $|\zeta_j| = \rho$, $|z_j| \le r < \rho$. Since f is bounded and measurable, we may multiply this series by f and integrate term by term. We obtain, for $|z_j| \le r < \rho$,

$$f(z) = \sum_{\alpha \in \mathbb{N}^n} c_\alpha z^\alpha, \quad c_\alpha = (2\pi i)^{-n} \int_{|\zeta_n|=\rho} d\zeta_n \cdots \int_{|\zeta_1|=\rho} (f(\zeta) \prod \zeta_j^{-\alpha_j - 1})\, d\zeta_1 ;$$

moreover, $|c_\alpha| \le M\rho^{-|\alpha|}$. Hence the above series converges uniformly to f for $|z_j| \le r$, and the lemma follows.

Lemma 4 (Baire's theorem). Let W be an open set in $\mathbb{R}^n$, and $\{W_p\}_{p=1,2,\dots}$ be a sequence of open dense subsets of W. Then $A = \bigcap_{p=1}^{\infty} W_p$ is dense in W.

Proof. Let $U_o \subset W$ be a nonempty open subset, and let $V_o \subset\subset U_o$, where V_o is a nonempty open set. Then $U_1 = V_o \cap W_1$ is a nonempty open set since W_1 is open and dense. Let V_1 be a nonempty open set, $V_1 \subset\subset U_1$. By induction on p, let $U_p = V_{p-1} \cap W_p$, and V_p be a nonempty open subset of U_p, $V_p \subset\subset U_p$. Then $\overline{V}_p \subset V_{p-1}$, hence $\bigcap V_p = \bigcap \overline{V}_p$. Since each $\overline{V}_p$ is a nonempty compact set and

$\overline{V}_p \subset \overline{V}_{p-1}$, we have $\bigcap V_p \neq \emptyset$. Clearly, $\bigcap V_p \subset (\bigcap W_p) \cap U_o$. Since U_o is arbitrary, the result follows.

Remark. The theorem is true when W is any complete metric space. We have only to replace the condition $V_p \subset\subset U_p$ by the conditions $\overline{V}_p \subset U_p$, diameter$(V_p) \to 0$ in the above proof.

Lemma 5. Let $\Omega = \{z \in \mathbb{C}^n \mid |z_j| < R\}$, let $\rho < R$, and let f be a function defined on Ω satisfying the conditions of Hartogs' theorem. Suppose that, for fixed a_n, $|a_n| < R$, the function $(z_1, \ldots, z_{n-1}) \mapsto f(z_1, \ldots, z_{n-1}, a_n)$ is holomorphic for $|z_j| < R$, $j = 1, \ldots, n-1$, and let $M(z') = \sup_{|a_n| \leq \rho} |f(z_1, \ldots, z_{n-1}, a_n)|$, $z' = (z_1, \ldots, z_{n-1})$. Then, there is an open dense set $U' \subset \Omega'$, $\Omega' = \{z' \in \mathbb{C}^{n-1} \mid |z_j'| < R, \; j = 1, \ldots, n-1\}$, such that $M(z')$ is bounded on any compact subset of U'.

Proof. Let V' be any nonempty open subset of Ω', and let $V_k' = \{z' \in V' \mid M(z') \leq k\}$, $k = 1, \ldots$. Then $\bigcup V_k' = V'$ and V_k' is closed in V': for each a_n, $|a_n| \leq \rho$, the set

$$\{z' \in V' \mid |f(z', a_n)| \leq k\}$$

is closed in V' since $z' \mapsto f(z', a_n)$ is holomorphic by assumption; V_k' is the intersection of these sets when a_n runs over the disc $|a_n| \leq \rho$. Hence $\bigcap (V' - V_k') = \emptyset$. By Lemma 4, at least one $V' - V_k'$ is not dense, so that V_{k_o}' contains a nonempty open set W' (for some k_o). Clearly $M(z') \leq k_o$, $z' \in W'$. Consider $U' = \bigcup W'$ when V' runs over all nonempty open subsets of Ω'. U' intersects each V', so is dense in Ω'. Moreover, any compact subset K' of U' is contained in the union of finitely many W', so that $M(z')$ is bounded for $z' \in K'$.

We can now prove Hartogs' theorem.

Proof of Hartogs' theorem. The theorem is trivial if $n = 1$. We assume, by induction, that the theorem has been proved for open sets in $\mathbb{C}^{n-1}$. Let $\Omega = \{z \in \mathbb{C}^n \,|\, |z_j| < R\} = \Omega' \times D$, $\Omega' = \{z' \in \mathbb{C}^{n-1} \,|\, |z_j| < R,\ j = 1, \ldots, n-1\}$, $D = \{z_n \in \mathbb{C} \,|\, |z_n| < R\}$. Let $\rho < R$, $\delta < R - \rho$ and U' be an open dense subset of Ω' such that f is bounded on $K' \times K_\rho$, $K_\rho = \{z \in \mathbb{C} \,|\, |z| \leq \rho\}$ for any compact $K' \subset U'$ (Lemma 5). Let $a' \in U'$, $|a'| < \delta$. Then f is holomorphic and bounded on the set $D' \times D_n$, $D' = \{z' \in \mathbb{C}^{n-1} \,|\, |z' - a'| < \eta\}$, $D_n = \{z \in \mathbb{C} \,|\, |z| < \rho\}$ if η is small enough (Lemma 3). Hence

$$\text{(S)} \qquad f(z) = \sum_{\alpha \in \mathbb{N}^{n-1}} a_\alpha(z_n)(z_1 - a_1)^{\alpha_1} \cdots (z_{n-1} - a_{n-1})^{\alpha_{n-1}},$$

the series converges uniformly on compact subsets of $D' \times D_n$, and the a_α are holomorphic on D_n.

By Cauchy's inequalites (Chapter 1, Proposition 3),

$$|a_\alpha(z_n)| \leq M\eta^{-|\alpha|}, \quad M = \sup_{z \in D' \times D_n} |f(z)|, \quad \forall \alpha \in \mathbb{N}^{n-1}.$$

On the other hand, for $|z_n| < R$, the function $(z_1, \ldots, z_{n-1}) \to f(z)$ is holomorphic on Ω' so that the series (S) converges uniformly on compact subsets of $\{z' \in \mathbb{C}^n \,|\, |z_j| < R - \delta\}$. Hence, in particular, for each z_n, there exists $A(z_n) > 0$ such that

$$|a_\alpha(z_n)| \leq A(z_n)\rho^{-|\alpha|}, \quad \forall \alpha \in \mathbb{N}^{n-1} \text{ since } \rho < R - \delta.$$

From these two relations, we conclude that

(a) $|a_\alpha(z_n)|^{1/|\alpha|}$ is bounded for $z_n \in D_n$, $\alpha \in \mathbb{N}^{n-1}$,

(b) $\overline{\lim\limits_{|\alpha| \to \infty}} |a_\alpha(z_n)|^{1/|\alpha|} \leq \frac{1}{\rho} \quad \forall z_n \in D_n$.

Hence, by Proposition 7, if $\varepsilon > 0$, and $r < \rho$ there exists $k_o > 0$ so that we have

$$|a_\alpha(z_n)| \le (\rho^{-1} + \varepsilon)^{|\alpha|} \quad \text{for } |z_n| \le r \text{ and } |\alpha| \ge k_o .$$

Hence, the series (S) converges uniformly for $|z_n| \le r$, $|z_1| \le \frac{\rho}{1+2\varepsilon\rho}, \ldots, |z_{n-1}| \le \frac{\rho}{1+2\varepsilon\rho}$. Since $\varepsilon > 0$ is arbitrary, and $r < \rho$ is arbitrary, it follows from Weierstrass' theorem (chapter 1, Proposition 5) that f is holomorphic on $\{z \in \mathbb{C}^n \mid |z_1| < \rho, \ldots, |z_n| < \rho\}$. The theorem is proved.

We prove next a proposition concerning subharmonic functions which we shall use in chapter 4.

Proposition 8. Let Ω be a connected open set in $\mathbb{C}$ and s a subharmonic function $\not\equiv -\infty$ on Ω. Let $A = \{x \in \Omega \mid s(x) = -\infty\}$ and suppose that A is closed. If u is a continuous function on Ω, subharmonic on $\Omega - A$, then u is subharmonic on Ω.

Proof. Since s is bounded above on compact subsets of Ω, we may suppose, replacing s by $s - c$, that $s \le 0$ on Ω. Let $\varepsilon > 0$. Then $u + \varepsilon s$ is subharmonic on Ω. In fact, condition (*) of Proposition 4 is trivially satisfied at points of A, (since $u + \varepsilon s = -\infty$ at these points) and outside, (*) is satisfied since $u + \varepsilon s$ is sub harmonic on $\Omega - A$. If $a \in \Omega$ and $\{z \mid |z-a| \le R\} \subset \Omega$ and $h(z) = \int_0^{2\pi} P_{a,R}(z,\theta) u(a + Re^{i\theta}) d\theta$, then by Proposition 3, $u + \varepsilon s \le h$ for $|z-a| = R$ (since $s \le 0$). Hence $u + \varepsilon s \le h$ on $|z-a| \le R$. Letting $\varepsilon \to 0$ we find that

$$u(z) \le h(z) \quad \text{if} \quad |z-a| \le R, \quad z \notin A.$$

But by Lemma 2, (a), $\Omega - A$ is dense in Ω. Since u and h are continuous, it follows that

$$u(z) \leq h(z) \quad \text{for} \quad |z-a| \leq R.$$

It follows that u is subharmonic.

Corollary. If s, A are as above, and u is continuous on Ω and harmonic on $\Omega - A$, then u is harmonic on Ω.

We may suppose that u is real valued. We have then only to apply Proposition 8 to u and $-u$.

Proposition 9. Let Ω, s, and A be as in Proposition 8. If then u is a bounded, continuous subharmonic function on $\Omega - A$, there is a subharmonic function v on Ω with $v|\Omega - A = u$.

Proof. We define v on A by

$$v(a) = \overline{\lim_{z \to a,\, z \notin A}}\; u(z) \;, \quad a \in A, \; v(z) = u(z) \text{ if } z \notin A.$$

Let $R > 0$ be so that $\{z \,|\, |z-a| \leq R\} \subset \Omega$. By Lemma 2, (b), the set $\{\theta \in [0, 2\pi] \,|\, s(a + Re^{i\theta}) = -\infty\}$ is of measure 0. Define h by

$$h(z) = \int_0^{2\pi} P_{a,R}(z, \theta) u(a + Re^{i\theta}) d\theta = \int_0^{2\pi} P_{a,R}(z, \theta) v(a + Re^{i\theta}) \, d\theta \,.$$

Then, as in the proof of Proposition 8, h is harmonic in $|z-a| < R$, and $v - h + \varepsilon s$ is subharmonic in $|z-a| < R$. Since $\overline{\lim_{z \to \zeta}} (v - h + \varepsilon s)(\zeta) \leq 0$ if $|\zeta - z| = R$, it follows that $v - h + \varepsilon s \leq 0$ on $|z-a| < R$. Hence letting $\varepsilon \to 0$

$$u(z) \leq h(z) \quad \text{for} \quad |z-a| < R, \quad z \notin A \,.$$

It follows from the definition of v that $v(z) \leq h(z)$, $|z-a| < R$, and v is thus subharmonic.

Corollary. If s, A are as above, any bounded harmonic function on $\Omega - A$ has a unique harmonic extension to Ω.

These results are merely special cases of much more precise results known in potential theory. For the theory of subharmonic functions, see [25]. Hartogs' theorm is proved in [13]. See also [17] for a presentation of this proof.

We have developed the theory of subharmonic functions to a point at which Radó's theorem (chapter 4, Theorem 1) and a generalization needed in chapter 5 (proof of Theorem 2 in chapter 5) become rather obvious.

4

HARTOGS' THEOREM ON THE SINGULARITIES OF HOLOMORPHIC FUNCTIONS

The object of this chapter is to prove a theorem due to Hartogs to the effect that the set of singularities of a holomorphic function tends to be an analytic set.

<u>Definition 1</u>. Let Ω be an open set in $\mathbb{C}^n$ and $A \subset \Omega$. We say that A is an analytic set if, for all $a \in \Omega$, there exists a neighborhood U of a and finitely many holomorphic functions $f_1, \ldots, f_p$ in U such that $A \cap U = \{f_1(z) = \ldots = f_p(z) = 0\}$.

<u>Proposition 1</u>. Let Ω be connected, A an analytic subset of Ω. Then we have the following.

(i) A is closed in Ω.

(ii) If $A \neq \Omega$, $\Omega - A$ is dense in Ω.

(iii) $\Omega - A$ is connected.

<u>Proof.</u> (i) By definition, any $a \in \Omega$ has a neighborhood U such that $A \cap U$ is closed in U. This implies that A is closed.

(ii) Suppose (ii) is false. Then $B = \overset{\circ}{A} \neq \emptyset$. We shall prove that B is closed in Ω; since B is open and Ω connected, this will prove that $A = \Omega$, a contradiction.

Let $a \in \overline{B}$, U be an open connected neighborhood of a in Ω, and let $f_1, \ldots, f_p$ be holomorphic in Ω and such that

$A \cap U = \{x \in U \mid f_1(x) = \ldots = f_p(x) = 0\}$. Then each f_j vanishes on $B \cap U$. Hence by the principle of analytic continuation, each $f_j \equiv 0$ on U, so that $U \subset A$; hence $U \subset \overset{\circ}{A} = B$; in particular, $a \in B$, so that $B = \overline{B}$.

(iii) It is enough to prove the following:

(*) Any $a \in \Omega$ has a connected neighborhood U such that $U - A$ is connected.

In fact, if (*) is proved, and $\Omega - A = U_1 \cup U_2$ where the U_j are open, nonempty, and disjoint, we have, by (ii), $\Omega = \overline{\Omega - A} = \overline{U}_1 \cup \overline{U}_2$. Since Ω is connected, $\overline{U}_1 \cap \overline{U}_2 \neq \emptyset$. Let U be a neighborhood of a such that $U - A$ is connected. Then $U - A = (U_1 - A) \cup (U_2 - A)$ cannot be connected unless one of U_1, U_2 is contained in A which is impossible by (ii).

To prove (*), let U be a convex neighborhood of a in which there exist holomorphic functions $f_1, \ldots, f_p$ with $U \cap A = \{x \in U \mid f_1(x) = \ldots = f_p(x) = 0\}$. Let $x_o, x_1 \in U - A$, and $V = \{\lambda \in \mathbb{C} \mid \lambda x_o + (1-\lambda)x_1 \in U\}$. V is a convex set in $\mathbb{C}$ and at least one $g_j(\lambda) = f_j(\lambda x_o + (1-\lambda)x_1) \not\equiv 0$ on V. Hence $A' = \{\lambda \in C \mid \lambda x_o + (1-\lambda)x_1 \in U \cap A\}$ is discrete in V. Hence $V - A'$ is connected and $0, 1 \in V - A'$. If $t \mapsto \gamma(t)$ is an arc in $V - A'$ connecting 0 to 1, $t \mapsto \gamma(t)x_o + (1-\gamma(t))x_1$ is an arc in $U - A$ connecting x_1 to x_o.

Proposition 2 (Riemann's continuation theorem). Let Ω be an open set in $\mathbb{C}^n$ and A an analytic set such that $\Omega - A$ is dense. Let f be holomorphic on $\Omega - A$ and suppose that for any $a \in A$, there exists a neighborhood U of a in Ω such that $f \mid U-A$ is bounded. Then there is a a unique holomorphic function F on Ω with $F \mid \Omega - A = f$.

Proof. The uniqueness is obvious.

We first prove the theorem when $n = 1$. We have only to prove that if f is bounded and holomorphic in $\{0 < |z| < \rho\}$, then f can be continued to the disc $\{|z| < \rho\}$. Now

$$f(z) = \sum_{-\infty}^{\infty} a_\nu z^\nu \quad , \quad a_\nu = \frac{1}{2\pi i} \int_{|z|=r} f(z) z^{-\nu-1}\, dz \quad \text{for any } 0 < r < \rho.$$

For $\nu < 0$, this integral $\to 0$ as $r \to 0$ since f is bounded. Since a is independent of r, $a_\nu = 0$ for $\nu < 0$ and the result follows.

In the general case, let $a \in A$, and let V be a connected neighborhood of a so that $f|V-A$ and in which there are holomorphic functions $f_1, \ldots, f_p$ with $A \cap V = \{x \in V | f_1(x) = \ldots = f_p(x) = 0\}$. We may suppose that $h \equiv f_1 \not\equiv 0$. By making a linear change of coordinates in $\mathbf{C}^n$, we may suppose that $a = 0$ and that $h(0, \ldots, 0, z_n) \not\equiv 0$ in a neighborhood of $z_n = 0$. There then exists $\delta > 0$ so that $h(0, \ldots, 0, z_n) \neq 0$ for $0 < |z_n| \leq \delta$. We denote $(z_1, \ldots, z_{n-1})$ by z'. Let $\varepsilon > 0$ be such that $h(z', z_n) \neq 0$ for $|z'| \leq \varepsilon$ and $|z_n| = \delta$. Consider the function

$$g(z) = \frac{1}{2\pi i} \int_{|t|=\delta} \frac{f(z', t)}{t - z_n}\, dt \ ;$$

note that for $|z'| \leq \varepsilon$, $|t| = \delta$, the point $(z', t) \in V-A$ and so g is holomorphic for $|z'| < \varepsilon$, $|z_n| < \delta$. (Note after Corollary 2 to Proposition 2 in chapter 1.) Moreover, for fixed z', $|z'| < \varepsilon$, the function $t \mapsto f(z', t)$ admits a holomorphic extension to the disc $|t| < \delta$ by the case $n = 1$ of the theorem (since $h(z', z_n) \neq 0$ for $|z_n| = \delta$, so that $z_n \mapsto h(z', z_n)$ has only finitely many zeros in $|z_n| < \delta$). Hence, by Cauchy's formula,

$$g(z) = f(z) \text{ if } z \in V-A, \quad |z'| < \varepsilon, \quad |z_n| < \delta.$$

Hence f can be continued to a neighborhood of any point of A. The result now follows since the extension is unique.

Theorem 1 (Radó's theorem). Let U, Ω be two open sets in $\mathbb{C}^n$, $U \subset \Omega$, and let f be holomorphic in U. Suppose that for every point $a \in (\partial U) \cap \Omega$ we have

$$\lim_{z \to a,\, z \in U} f(z) = 0.$$

Then the function defined by

$$F(z) = \begin{cases} f(z) & \text{if } z \in U \\ 0 & \text{on } \Omega - U \end{cases}$$

is holomorphic in Ω.

This is equivalent to the following.

Theorem 1'. Let f be a continuous function on the open set $\Omega \subset \mathbb{C}^n$ and let $U = \{z \in \Omega \mid f(z) \neq 0\}$. Suppose that f is holomorphic on U. Then f is holomorphic on Ω.

Theorem 1 $\Rightarrow$ Theorem 1'. Given f as in Theorem 1', the function F defined in Theorem 1 is $= f$.

Theorem 1' $\Rightarrow$ Theorem 1. It is enough to check that if f is given on U and tends to 0 at $(\partial U) \cap \Omega$, then F is continuous. This is obvious.

Proof of Theorem 1'. It is enough to prove the theorem when $n = 1$. For, assuming the theorem in the case $n = 1$, if we consider $a_1, \ldots, a_{n-1} \in \mathbb{C}$, then the function $z_j \mapsto f(a_1, \ldots, a_{j-1}, z_j, a_j, \ldots, a_n)$ satisfies the conditions of Theorem 1' in an open set in $\mathbb{C}$ and so is holomorphic. Hence, by chapter 3, Lemma 3, f is holomorphic on Ω.

The theorem being local, we may suppose that Ω is connected. Moreover, we may suppose that $f \not\equiv 0$. Let $s(z) = \log|f(z)|$. Then s is subharmonic in Ω: Condition (*) of chapter 3, Proposition 4 is trivially verified at points of $A = \{z \in \Omega | F(z) = 0\} = \{z \in \Omega | s(z) = -\infty\}$; on Ω-A, s is harmonic, and condition (*) is again verified. Moreover, $f|\Omega - A$ is holomorphic, hence harmonic. By chapter 3, Corollary to Proposition 8, f is harmonic on Ω, in particular continuously differentiable. Since $\frac{\partial f}{\partial \bar{z}} = 0$ on $\Omega - A$ and $\Omega - A$ is dense (chapter 3, Lemma 2, (a)), we have $\frac{\partial f}{\partial \bar{z}} = 0$ on Ω, so that f is holomorphic.

The idea of the above proof is due to H. Cartan [9]. It is a simple matter to eliminate all reference to subharmonic functions and to prove directly, using chapter 3, Proposition 3, the special case of chapter 3, Proposition 8 needed. This has been pointed out by E. Heinz [15].

Definition 2. (a) A function defined on a subset $S \subset \mathbb{C}^n$ is called holomorphic on S if it is the restriction to S of a function holomorphic on an open set $U \supset S$.

(b) Let Ω be open, $A \subset \Omega$. If f is holomorphic on Ω-A, we say that f is singular at $a \in A$ if there is no holomorphic function in a neighborhood U of a whose restriction to $U - A$ is $f|U - A$.

These definitions are not the best ones available, but we give them simply for convenience in stating the main theorem of this chapter.

Theorem 2 (Hartogs' continuity theorem). Let Ω be a connected open set in $\mathbb{C}^n$ and let $Q = \{w \in \mathbb{C} \mid r < |w| < R\}$, $0 \leq r < R$. Let f be holomorphic on $\Omega \times Q$. Suppose that there is a point $a \in \Omega$ such that f can be continued holomorphically to a neighborhood of $\{a\} \times D$, $D = \{w \in \mathbb{C} \mid |w| < R\}$. Then f can be continued holomorphically to $\Omega \times D$.

Proof. Let $f(z,w) = \sum_{\nu=-\infty}^{\infty} a_\nu(z)w^\nu$ be the Laurent expansion of f which converges uniformally on compact subsets of $\Omega \times Q$. Then the a_ν are holomorphic on Ω. Now, if $r < \rho < R$, there is $\varepsilon > 0$ so that $f(z,w)$ can be extended to a holomorphic function $F(z,w)$ on $\{|z-a| < \varepsilon, |w| < \rho\}$ (since f can be extended to a neighborhood of $\{a\} \times D$). Hence, for $|z-a| < \varepsilon$, the a_ν with $\nu < 0$, are $= 0$. By the principle of analytic continuation, $a_\nu \equiv 0$ on Ω if $\nu < 0$, so that

$$f(z,w) = \sum_{\nu=0}^{\infty} a_\nu(z)w^\nu \quad , \ (z,w) \in \Omega \times Q .$$

This latter series then converges uniformly on compact subsets of $\Omega \times D$, and the result is proved.

Corollary. Let f be a function holomorphic on the set $z = 0$, $0 < |w| < R$ in $\mathbb{C}^n \times \mathbb{C}$. Suppose that $(0,0)$ is a singularity of f. Let $\delta > 0$ be sufficiently small. Then, there exists $\varepsilon > 0$ so that for any z_0, $|z_0| < \varepsilon$, f cannot be extended holomorphically to any neighborhood of $\{z_0\} \times D$, $D = \{|w| < \delta\}$.

Proof. Let $\varepsilon > 0$ be so chosen that f is holomorphic on the set $|z| < \varepsilon$, $r_0 < |w| < r_1$, $r_0, r_1 < \delta$. The corollary follows from Theorem 2.

Our next main theorem is again due to Hartogs.

Theorem 3. Let Ω be a connected open set in $\mathbb{C}^n$ and $\varphi:\Omega \to \mathbb{C}$ a map. Suppose that $|\varphi(z)| < R$ for all $z \in \Omega$ and let $U = \Omega \times D$, $D = \{w \in \mathbb{C} \mid |w| < R\}$. Let $\Gamma = \{(z,w) \in U) \mid \varphi(z) = w\}$. Suppose that f is a function holomorphic on $U - \Gamma$ which is singular at every point of Γ. Then φ is a holomorphic function of z.

The proof requires several preliminaries.

Lemma 1. Let Ω, φ, f, R, D be as in Theorem 3. Then φ is continuous.

Proof. Let $z_0 \in \Omega$, $\varphi(z_0) = w_0$. Let $\varepsilon > 0$. By the corollary to Theorem 2, there is $\delta > 0$ so that f cannot be extended holomorphically to the set $\{z\} \times \{|w-w_0| < \varepsilon\}$ for $|z-z_0| < \delta$. Since, for $|z-z_0| < \delta$ f is holomorphic on $\{z\} \times (D - \{\varphi(z)\})$, it follows that $|\varphi(z) - w_0| < \varepsilon$, so that φ is continuous.

Let Ω be an open set in $\mathbb{C}$ and $\{f_\nu\}$, $\nu = 0, 1, \ldots$ a sequence of holomorphic functions on Ω. We consider the series

(H) $$\sum_{\nu=0}^{\infty} f_\nu(z) w^\nu \quad ,$$

and suppose that it converges uniformly on compact subsets of a neighborhood of $\Omega \times \{0\}$.

Definition. The Hartogs radius of the above series is the function $R:\Omega \to \mathbb{R}^+$ defined by

$$R(z_0) = \sup\{|w| \mid \sum_0^{\infty} f_\nu(z) w^\nu \text{ converges uniformly on a neighborhood of } (z_0, w)\} .$$

Since the series converges uniformly on a neighborhood of $\Omega \times \{0\}$, it is clear that $R(z_0) > 0$ for all $z_0 \in \Omega$.

Proposition 3. If $R(z) \not\equiv +\infty$, the function $-\log R(z)$ is subharmonic in Ω.

Proof. To see that $-\log R(z)$ is u.s.c., let $r < R(z_0)$. Then there is $\delta > 0$ such that the series converges uniformly in a neighborhood of the set $|z - z_0| \leq \delta$, $w = r$, hence on $|z - z_0| \leq \delta$, $|w| \leq r$. In particular $R(z) \geq r$ for $|z - z_0| < \delta$, and we have $\underline{\lim}_{r \to z_0} R(z) \geq R(z_0)$.

Let $K = \{ z \in \mathbb{C} \mid |z - z_0| \leq \rho\} \subset \Omega$. Let h be continuous on K and harmonic on $\overset{o}{K}$. Suppose that, for $|z-z_0| = \rho$,

$$\text{(I)} \qquad -\log R(z) \leq h(z) .$$

We have to show that (I) holds for $|z-z_0| < \rho$. Since the series (H) converges for $|z-z_0| = \rho$, $|w| < R(z)$, we have

$$\overline{\lim_{\nu \to \infty}} |f_\nu(z)|^{1/\nu} \leq \frac{1}{R(z)} \leq e^{h(z)} \quad \text{for } |z-z_0| = \rho .$$

On the other hand, $R(z)$ is bounded below on K, hence (H) converges uniformly on a set of the form $K \times \{|w| \leq \eta\}$, $\eta > 0$. In particular,

$$|f_\nu(z)|\eta^\nu \leq M , \quad \forall z \in K, \quad \forall \nu = 0, 1, \ldots ,$$

where M is a constant.

Hence by the remark following Proposition 7, chapter 3, we have

$$\overline{\lim_{\nu \to \infty}} \alpha_\nu(r) \leq 0,$$

where $\alpha_\nu(r) = \sup_{|z-z_0| \leq r} \{ \frac{1}{\nu} \log |f_\nu(z)| - h(z)|\}$, $0 < r < \rho$.

Hence, for $\varepsilon > 0$, $|f_\nu(z)|^{1/\nu} \leq e^{h(z)+\varepsilon}$ uniformly for $|z-z_0| \leq r$ and $\nu > \nu_0(\varepsilon)$. Hence the series (H) converges uniformly in the neighborhood of (z, w) for $|z-z_0| < \rho$, $|w| < e^{-h(z)-\varepsilon}$. Hence $R(z) \geq e^{-h(z)}$ for $|z-z_0| < \rho$, which proves (I) for these z.

We now turn to the proof of Theorem 3 in the case $n = 1$. We may suppose that $0 \in \Omega \subset \mathbb{C}$ and that $\varphi(0) = 0$. We have to show that φ is holomorphic in the neighborhood of $z = 0$. Let $\frac{R}{2} > \eta > 0$ and $\varphi > 0$ be so that $|\varphi(z)| < \eta$ for $|z| < \delta$ (Lemma 1). Let $w_o \in \mathbb{C}$, $\eta < |w_o| < R-\eta$. Then f is holomorphic in the neighborhood of (z,w_o) for $|z| < \delta$. Let

$$(H_o) \qquad f(z,w) = \sum_{k=0}^{\infty} f_k(z)(w-w_o)^k .$$

The above series converges uniformly in the neighborhood of (z,w) for $|w-w_o| < |\varphi(z) - w_o|$ since f is holomorphic in a neighborhood of such a (z,w) (note that $|w_o| < R-\eta$, so that $|\varphi(z) - w_o| < R$). Hence, the Hartogs radius associated with such a series, $R(z)$, satisfies the inequality $R(z) \geq |\varphi(z) - w_o|$. On the other hand, if $R(z_o) > |\varphi(z_o) - w_o|$ for some z_o, $|z_o| < \delta$, we would have, since R is semicontinuous, $R(z) > \rho$ for all z near z_o, where ρ is a number with $|\varphi(z_o) - w_o| < \rho < R(z_o)$. This would imply the uniform convergence of the series for z near z_o, $|w| < \rho$; in particular, f would have an analytic extension to a neighborhood of the point $(z_o, \varphi(z_o))$ since $|\varphi(z_o) - w_o| < \rho$, which is impossible by assumption. Thus we have

Lemma 2. The Hartogs radius R of the series (H_o) is given by

$$R(z) = |\varphi(z) - w_o| \quad \text{for } |z| < \delta .$$

Proposition 4. If δ, η and φ are as above, the function

$$z \mapsto \log|\varphi(z) - w_o|$$

is harmonic in $|z| < \delta$ for any w_o with $\eta < |w_o| < R-\eta$.

Proof. Let $u(z, w_o) = -\log|\varphi(z) - w_o|$. By Proposition 3 and Lemma 2, $u(z, w_o)$ is a subharmonic function of z in $|z| < \delta$. Hence, for $\rho < \delta - |z|$, we have

$$\text{(II)} \qquad u(z, w_o) \leq \frac{1}{2\pi} \int_0^{2\pi} u(z + \rho e^{i\theta}, w_o) d\theta \ .$$

We set $w_o = re^{i\varphi}$ in this inequality; for fixed r, $\eta < r < R-\eta$, we have (II) for $0 \leq \varphi \leq 2\pi$. We integrate with respect to φ from 0 to 2π. Now, $\int_0^{2\pi} u(\zeta, re^{i\varphi}) d\varphi = -\int_0^{2\pi} \log|\varphi(\zeta) - re^{i\varphi}| d\varphi = 2\pi \log \frac{1}{r}$ for $|\zeta| < \delta$ since $|\varphi(\zeta)| < \eta$ and $\int_0^{2\pi} \log|\alpha - re^{i\varphi}| d\varphi = 2\pi \log r$ for $|\alpha| < r$. [To prove this latter formula, it is enough to remark that $\log|\alpha - re^{i\varphi}| = \log|\alpha e^{-i\varphi} - r| = \log|\overline{\alpha} e^{i\varphi} - r|$, and $\log(z - r)$ is holomorphic in $|z| < r$, so that we can use chapter 3, Proposition 3, Corollary 1.] This gives

$$\int_0^{2\pi} u(z, re^{i\varphi}) d\varphi = 2\pi \log \frac{1}{r} = \frac{1}{2\pi} \int_0^{2\pi} d\varphi \int_0^{2\pi} u(z+\rho e^{i\theta}, re^{i\varphi}) d\theta \ .$$

This and (II) imply, since u is continuous, that

$$u(z, w_o) = \frac{1}{2\pi} \int_0^{2\pi} u(z + \rho e^{i\theta}, w_o) d\theta \quad \text{for } \eta < |w_o| < R-\eta, \ \rho < \delta - |z| .$$

By Corollary 7 to Proposition 4 in chapter 3, this proves the result.

Lemma 3. If $\log|\varphi(z) - w|$ is harmonic in z for all w, $\eta < |w| < \rho - \eta$, then φ or $\overline{\varphi}$ is a holomorphic function of z.

Proof. Since $\log|\varphi(z) - w|$ is harmonic, it is a real analytic function, hence C^∞; hence so is $|\varphi(z) - w|^2 = \exp(2 \log|\varphi(z) - w|)$, and $\psi(z, w) = |\varphi(z)|^2 - w\varphi(z) - \overline{w}\overline{\varphi}(z)$ is C^∞ for $w = re^{i\lambda}$, $\eta < r < R-\eta$, $0 \leq \lambda \leq 2\pi$. Hence $w\varphi + \overline{w}\overline{\varphi} = \frac{1}{2}(\psi(z, -w) - \psi(z, w))$ is C^∞. Taking $w = r$, $w = ir$, we find that $\varphi + \overline{\varphi}$ and $\varphi - \overline{\varphi}$, hence also φ, are C^∞.

To prove our result, we have now to show that either $\partial\varphi/\partial\bar{z}$ or $\partial\bar{\varphi}/\partial\bar{z}$ is identically 0. Now

$$0 = \frac{\partial^2}{\partial z\partial\bar{z}} \log(\varphi(z)-w)(\overline{\varphi(z)}-\bar{w}) = (\varphi-w)^{-1}\frac{\partial^2\varphi}{\partial z\partial\bar{z}} - (\varphi-w)^{-2}\frac{\partial\varphi}{\partial z}\cdot\frac{\partial\varphi}{\partial\bar{z}} + (\bar{\varphi}-\bar{w})^{-1}\frac{\partial^2\bar{\varphi}}{\partial z\partial\bar{z}} - (\bar{\varphi}-\bar{w})^{-2}\frac{\partial\bar{\varphi}}{\partial z}\cdot\frac{\partial\bar{\varphi}}{\partial\bar{z}},$$

for all w in a nonempty open set. Multiplying by $(\varphi-w)^{-2}(\bar{\varphi}-\bar{w})^2$ and equating powers of w and $\bar{w}$, we find that

$$\frac{\partial^2\varphi}{\partial z\partial\bar{z}} = 0, \quad \frac{\partial\varphi}{\partial z}\frac{\partial\varphi}{\partial\bar{z}} = 0.$$

The first relation implies that φ is harmonic, hence real-analytic, and the second that either $\frac{\partial\varphi}{\partial z}$ or $\frac{\partial\varphi}{\partial\bar{z}}$ vanishes on a nonempty open set. Hence, by the principle of analytic continuation, either $\frac{\partial\varphi}{\partial\bar{z}}$ or $\frac{\partial\bar{\varphi}}{\partial\bar{z}} = (\overline{\frac{\partial\varphi}{\partial z}})$ is identically 0, and the lemma is proved.

Corollary. Under the conditions of Theorem 3, if $n = 1$, then either φ or $\bar{\varphi}$ is holomorphic.

Proof of Theorem 3 when n = 1. We write $\omega = w + z$, $\zeta = z$. Then, for $|\zeta|, |\omega|$ sufficiently small, the set $\{(\zeta,\omega) \mid \omega = \varphi(\zeta) + \zeta\}$ satisfies the hypotheses of Theorem 3, $n = 1$ with respect to the function $f(\zeta, \omega-\zeta)$. Hence, by the above corollary, in a sufficiently small neighborhood of $\zeta=0$, $\varphi(\zeta)$ or $\overline{\varphi(\zeta)} + \bar{\zeta}$ is holomorphic. If φ is not holomorphic, then $\bar{\varphi}(\zeta)$ and $\bar{\varphi}(\zeta) + \bar{\zeta}$ are holomorphic, hence so is $\bar{\zeta}$, which is absurd. Hence φ is holomorphic at 0, and the result is proved.

Proof of Theorem 3 in the general case. Let $a \in \Omega$ and P be a sufficiently small polydisc with center a, $P \subset \Omega$. Let $z_o \in P$. We shall prove that the function $\lambda \to \varphi(a + \lambda(z_o - a))$ is holomorphic in the convex set $D = \{\lambda \in \mathbf{C} \mid a + \lambda(z_o - a) \in P\}$. Since the limit of holomorphic functions is again holomorphic, it suffices to prove this for z_o in a dense subset $S \subset P$. We may suppose that $a = 0$, $\varphi(0) = 0$, that $P = \{|z| < \delta\}$, and that $|\varphi(z)| < \eta$ for $|z| < \delta$, η small. Then f is holomorphic on $\{(z, w) \in \mathbb{C}^{n+1} \,|\, |z| < \delta, \ \eta < |w| < R\}$, so that we may expand in a Laurent series:

$$f(z, w) = \sum_{\nu = -\infty}^{\infty} f_\nu(z) w^\nu$$

Since f is singular at $(0, 0)$, there is $\nu_o < 0$ so that $f_{\nu_o} \not\equiv 0$ on $P = \{|z| < \delta\}$. Let $S = \{z \in P \,|\, f_{\nu_o}(z) \neq 0\}$. The function

$$g(\lambda, w) = f(\lambda z_o, w) \quad , \quad z_o \in S$$

is holomorphic on the set $\{(\lambda, w) \,|\, \lambda \in D, \ w \neq \varphi(\lambda z_o)\}$ and is moreover singular at some point (λ, w_o) for each λ (since otherwise $f_{\nu_o}(\mu z_o) \equiv 0$ for μ in a neighborhood of some λ, hence $f_{\nu_o}(z_o) = 0$ by the principle of analytic continuation, a contradiction). Hence $g(\lambda, w)$ is singular precisely on the set $w = \varphi(\lambda z_o)$, and the result follows from the case $n = 1$ proved above.

That φ is holomorphic on Ω follows now from chapter 3, Lemma 3.

<u>Proposition 5.</u> Let Ω be a connected set in $\mathbb{C}^n$, $D = \{w \in \mathbb{C} \,|\, |w| < R\}$ and let K be a compact subset of D. Let $A \subset \Omega \times K$ and suppose that there exists $p > 0$ so that, for all $z \in \Omega$,

$$A_z = \{w \in D \,|\, (z, w) \in A\}$$

is a finite set containing at most p elements. Suppose that there is a holomorphic function f on $\Omega \times D - A$ which is singular at every point of A. Then, there are holomorphic functions $\varphi_1, \ldots, \varphi_q$ on Ω, $q \leq p$, so that

$$A = \{(z,w) \in \Omega \times D \mid w^q + \varphi_1(z)w^{q-1} + \ldots + \varphi_q(z) = 0\} \; ;$$

in particular, A is an analytic set.

Proof. Let q be the maximum number of elements in A_z when z runs over Ω, and let U be the set of $z \in \Omega$ so that A_z contains q elements. We claim that U is open. Let $a \in U$ and $\{w_1, \ldots, w_q\} = A_a$. Let $\varepsilon < \frac{1}{2} \max_{i \neq j} |w_i - w_j|$. It is enough to prove that, if z is near a, $A_z \cap \{w \in \mathbb{C} \mid |w - w_i| < \varepsilon\} \neq \emptyset$ for $i = 1, \ldots, q$. Since A_z is, by assumption, the set of singularities of the form (z,w) of f, this follows at once from Theorem 2.

For $z \in U$, let $A_z = \{w_1(z), \ldots, w_q(z)\}$, and let

$$g(z) = \prod_{i<j} (w_i(z) - w_j(z))^2 \, .$$

We claim that g is holomorphic on U. To prove this, let $a \in U$ and z be near a. Let $\varepsilon < \frac{1}{2} \max_{i = j} |w_i(a) - w_j(a)|$, and for each i and z near a, let $w_i(z)$ be that element of A_z with $|w_i(z) - w_i(a)| < \varepsilon$. For z near a, we can apply Theorem 3 to $w_i(z)$; hence $w_i(z)$ is holomorphic in z for z near a, and it follows that g is holomorphic.

Let $a_\nu \in U$, $a_\nu \to a \in (\partial U) \cap \Omega$. Since, by assumption $w_i(a_\nu) \in K$, there is a subsequence $\{\nu_k\}$ so that $w_i(a_\nu) \to w_i \in K$, $i = 1, 2, \ldots, q$. Since A is closed (see remark after Definition 1), $(a, w_i) \in A$, $i = 1, \ldots, q$. Since $a \notin U$, we conclude that $w_i = w_j$ for some $i \neq j$. It follows, since each $w_i(z)$ is obviously bounded on U (note that $w_i(z) \in K$), that $g(a_{\nu_k}) \to 0$ as $k \to \infty$. Since this is true for any sequence $a_\nu \to a$, it follows that $g(a_\nu) \to 0$ as $\nu \to \infty$. Hence, by Radó's theorem (Theorem 1), the function h on Ω defined by

$$h(z) = \begin{cases} g(z) & , \ z \in U \\ 0 & , \ z \in \Omega - U \end{cases}$$

is holomorphic on Ω. Moreover, $\Omega - U = \{z \in \Omega \mid h(z) = 0\}$, and so is an analytic set in Ω.

For $z \in U$, define

$$P'(z,w) = \prod_{i=1}^{q} (w - w_i(z)) = w^q + \varphi_1'(z)w^{q-1} + \ldots + \varphi_q'(z),$$

where $\varphi_k'(z) = (-1)^k \sum_{1 \leq i_1 < \ldots < i_k \leq q} w_{i_1}(z) \cdots w_{i_k}(z)$ is the k-th elementary symmetryc function in the $w_i(z)$. Since $w_i(z) \in K$, φ_k' is bounded on U. Further, the φ_k' are holomorphic on U (see proof that g is holomorphic given above). Hence, by Proposition 2, there are holomorphic functions φ_k on Ω, whose restrictions to U are the φ_k' above.

Let $P(z,w) = w^q + \varphi_1(z)w^{q-1} + \ldots + \varphi_q(z)$. If $(z,w) \in \Omega \times \mathbb{C}$ and $P(z,w) = 0$, then $w \in K$. Clearly, for $z \in U$, $A_z = \{w \in \mathbb{C} \mid P(z,w) = 0\}$. Since U is dense in Ω, we conclude that $A \subset \{(z,w) \in \Omega \times \mathbb{C} \mid P(z,w) = 0\}$. Conversely, if $(z,w) \in \Omega \times \mathbb{C}$ and $P(z,w) = 0$, then $w \in K \subset D$ and if $z_\nu \to z$, $z_\nu \in U$, there are $w_\nu \in \mathbb{C}$, $P(z_\nu, w_\nu) = 0$ so that $w_\nu \to w$. Since $(z_\nu, w_\nu) \in A_{z_\nu} \subset A$ and A is closed, $(z,w) \in A$, so that $A = \{(z,w) \in \Omega \times \mathbb{C} \mid P(z,w) = 0\}$.

<u>Theorem 4.</u> Let W be an open set in $\mathbb{C}^{n+1}$ and $A \subset W$ a subset such that

(a) For all $a \in W$ there is a neighborhood $\Omega \times D$, $\Omega \subset \mathbb{C}^n$, $D = \{ |w - a_{n+1}| < \rho \} \subset \mathbb{C}$, and $p > 0$ such that $\{w \in D \mid (z,w) \in A\}$ contains at most p points for any $z \in \Omega$.

(b) There is f holomorphic on $W - A$ which is singular at every point of A.

Then A is an analytic subset of W.

Proof. Let $a \in W$ and $\Omega \times D$ be as in (a). Consider $\{w \in D \mid (a', w) \in A\} = \{w_1, \ldots, w_q\}$, $a' = (a_1, \ldots, a_n)$, $a = (a_1, \ldots, a_{n+1})$. Let $0 < r < \rho$ be so that $\{a'\} \times \{w \mid |w - a_{n+1}| = r\} \cap A = \emptyset$. Then, if ε is small enough, the set $|z - a'| \leq \varepsilon$, $r - \varepsilon \leq |w - a_{n+1}| \leq r + \varepsilon$ does not meet A. If $P = \{z \in \mathbb{C}^n \mid |z - a'| < \varepsilon\}$, $D_0 = \{w \in \mathbb{C} \mid |w - a_{n+1}| < r + \varepsilon\}$, it follows that the projection on D_0 of $A \cap P \times D_0$ is contained in the compact set $\{w \mid |w - a_{n+1}| \leq r\}$ of D. By Proposition 5, there are holomorphic functions $\varphi_1, \ldots, \varphi_q$, $q \leq p$, on P with $A \cap P \times D_0 = \{(z, w) \in P \times D_0 \mid w^q + \varphi_1(z) w^{q-1} + \ldots + \varphi_q(z) = 0\}$. Clearly, then, A is an analytic set.

For the results on analytic sets, see [16], [20]. Hartogs proved Theorem 3 in [14].

A form of Radó's theorem occurs in [26]. Its usefulness in complex analysis was recognized by H. Behnke and K. Stein.

We have assumed in Proposition 5 that the number of points in the set A_z is bounded as a function of z. The extent to which this hypothesis is essential is unknown.

5

AUTOMORPHISMS OF BOUNDED DOMAINS

Let Ω be an open set in $\mathbb{C}^n$ and $f = (f_1, \ldots, f_m): \Omega \to \mathbb{C}^m$ a map. We say that f is holomorphic if each f_j is, $1 \leq j \leq m$.

Definition 1. Let Ω be a connected open set (domain) in $\mathbb{C}^n$. A holomorphic map $f: \Omega \to \Omega$ is called an (analytic) automorphism if there is a holomorphic map $g: \Omega \to \Omega$ with $g \circ f = \text{id.}$, $f \circ g = \text{id.}$ (id is the identity map of Ω.)

Remark. A theorem of Osgood, which we shall prove at the end of this chapter, implies that a bijective holomorphic map of Ω onto itself is an automorphism.

If D is a bounded domain in $\mathbb{C}^n$, we denote by $\mathrm{Aut}(D)$ the set of automorphisms of D. $\mathrm{Aut}(D)$ is a group relative to composition of maps maps. We introduce a topology on $\mathrm{Aut}(D)$ as follows.

Let $K \subset D$ be compact, U and open set contained in D. Then, the set $\{\sigma \in \mathrm{Aut}(D) \mid \sigma(K) \subset U\}$ runs over a basis of open sets for the topology of $\mathrm{Aut}(D)$ when K runs over the compact subsets of D, U over the open sets. This topology has a countable basis formed by the sets $\{\sigma \in \mathrm{Aut}(D) \mid \sigma(K_\nu) \subset U_\mu\}$ where $\{U_\mu\}_{\mu = 1, 2, \ldots}$ is a countable base of open sets in D and $\{K_\nu\}_{= 1, 2, \ldots}$ is a sequence of compact sets such that the $\{\overset{\circ}{K}_\nu\}$ form a basis of open sets for $D\}$.

A sequence $\{\sigma_\nu\} \subset \mathrm{Aut}(D)$ converges if and only if σ_ν converges uniformly on compact subsets of D to an element $\sigma \in \mathrm{Aut}(D)$. Moreover, $\mathrm{Aut}(D)$ is Hausdorff, and is a topological group [the map $(\sigma, \tau) \mapsto \sigma \circ \tau^{-1}$ is continuous].

Proposition 1 (H. Cartan). Let D be a bounded domain in $\mathbf{C}^n$ and $f: D \to D$ a holomorphic map. Suppose that there exists $a \in D$ with $f(a) = a$. Let the expansion of f in a series of homogeneous polynomials about a be of the form

$$f(z) = z + P_2(z-a) + \dots$$

where P_k is an n-tuple of homogeneous polynomials of degree k.

Then $f(z) \equiv z$.

Proof. We may suppose that $a = 0$. Suppose that D is contained in the polydisc $\{z \mid |z_j| < R\}$ (such an $R > 0$ exists since D is bounded). If F is a holomorphic map of D into itself, $F = (F_1, \dots, F_n)$, and $F(z) = \Sigma\, a_\alpha z^\alpha$ $(a_\alpha = (a_\alpha^1, \dots, a_\alpha^n) \in \mathbf{C}^n)$ is the Taylor expansion about 0, then, by Cauchy's inequalities (chapter 1, Proposition 3), we have $|a_\alpha| \leq R\rho^{-|\alpha|}$, $\rho > 0$ being such that $\{z \in \mathbf{C}^n \mid |z_j| < \rho\} \subset D$.

Consider now f^k, the k-th iterate of f defined by $f^1 = f$, $f^k = f \circ f^{k-1}$. Suppose that $f(z) \not\equiv z$. Let N be the smallest integer such that

$$f(z) = z + P_N(z) + \dots, \qquad P_N \not\equiv 0 .$$

One sees easily, by induction on k, that

$$f^k(z) = z + kP_N(z) + \dots .$$

Since f^k is a map of D into itself, our remark above shows that the coefficients of $kP_N(z)$ are $\leq R\rho^{-N}$ in absolute value. Since k is arbitrary, this implies $P_N \equiv 0$, a contradiction.

Definition 2. A bounded domain $D \subset \mathbb{C}^n$ is called a circular domain if $z \in D$ and $\theta \in \mathbb{R}$ implies that $e^{i\theta}z \in D$.

Proposition 2 (H. Cartan). Let D be a circular domain in $\mathbb{C}^n$ and $f = (f_1, \ldots, f_n) \in \mathrm{Aut}\, D$. Suppose that $0 \in D$ and that $f(0) = 0$. Then each f_j is linear, i.e.,

$$f_j(z) = a_{1j}z_1 + \ldots + a_{nj}z_n \ , \quad a_{ij} \in \mathbb{C}.$$

Proof. For any holomorphic map $g: D \to D$, we denote by $dg = (dg)_0$ the linear map of $\mathbb{C}^n$ into itself given by

$$(dg)(v_1, \ldots, v_n) = (w_1, \ldots, w_n)\ , \quad w_k = \sum_{j=1}^{n} \frac{\partial g_k}{\partial z_j}(0)\cdot v_j \ , \quad g = (g_1, \ldots, g_n).$$

Let $\theta \in \mathbb{R}$, and $k_\theta \in \mathrm{Aut}(D)$ the map $z \to e^{i\theta}z$. Then the inverse of k_θ is $k_{-\theta}$. Let φ be the inverse of f, and let

$$g = k_{-\theta} \circ \varphi \circ k_\theta \circ f\ .$$

Since k_θ and f, hence also $k_{-\theta}$ and φ leave the origin invariant, we find that

$$dg = dk_{-\theta} \circ d\varphi \circ dk_\theta \circ df\ .$$

Now, $dk_\theta = e^{i\theta}\mathrm{id.}$ is a linear map of $\mathbb{C}^n$ into itself which commutes with any linear map of $\mathbb{C}^n$. Hence $dg = (dk_{-\theta} \circ dk_\theta) \circ (d\varphi \circ df) = \mathrm{identity}$. Since clearly $dk_{-\theta} \circ dk_\theta = \mathrm{id.} = d\varphi \circ df$. Hence $g(0) = 0$ and $dg = \mathrm{id.}$ It follows that the expansion of g in a series of homogeneous polynomials about 0 has the form

$$g(z) = z + P_2(z) + \dots$$

so that, by Proposition 1, $g(z) \equiv z$. This can be written

$$k_\theta \circ f = f \circ k_\theta \quad \text{for all } \theta \in \mathbb{R} .$$

Hence, if $f = (f_1, \dots, f_n)$, we have $f_j(e^{i\theta} z) = e^{i\theta} f_j(z)$. If $f_j(z) = \sum_{\alpha \in \mathbb{N}^n} a_\alpha z^\alpha$, we deduce that

$$e^{i\theta} a_\alpha = e^{i|\alpha|\theta} a_\alpha \quad \text{for all } \theta \in \mathbb{R} .$$

This implies that $a_\alpha = 0$ if $|\alpha| \neq 1$, and the result follows.

Definition 3. A holomorphic map $f: D \to D'$, $D \subset \mathbb{C}^n$, $D' \subset \mathbb{C}^m$ is called proper if, for any compact $K' \subset D'$, the set $f^{-1}(K')$ is compact in D.

Remark. Trivially, an automorphism of a domain D is proper. Moreover, $f: D \to D'$ is proper if and only if for any sequence $\{z_\nu\} \subset D$ which has no limit point in D, the sequence $\{f(z_\nu)\}$ has no limit point in D'.

We shall now see how these results can be applied to determine all automorphisms of a polydisc.

Proposition 3. Let $D = \{z \in \mathbb{C}^n \mid |z_j| < 1\}$. Then, for any $f \in \mathrm{Aut}(D)$, there exists a permutation $p: (1, \dots, n) \to (1, \dots, n)$ of the integers from 1 to n, real numbers $\theta_1, \dots, \theta_n \in \mathbb{R}$, and complex numbers $\alpha_1, \dots, \alpha_n$, $|\alpha_j| < 1$, so that

$$f(z) = \left(e^{i\theta_1} \frac{z_{p(1)} - \alpha_1}{1 - \bar{\alpha}_1 z_{p(1)}}, \dots, e^{i\theta_n} \frac{z_{p(n)} - \alpha_n}{1 - \bar{\alpha}_n z_{p(n)}}\right).$$

Proof. Let $\sigma_\alpha(z) = \left(\frac{z_1 - \alpha_1}{1 - \bar{\alpha}_1 z_1}, \ldots, \frac{z_n - \alpha_n}{1 - \bar{\alpha}_n z_n}\right)$, $|\alpha_j| < 1$. Then $\sigma_\alpha \in \mathrm{Aut}(D)$ for $|\alpha_j| < 1$ and, for $z_0 \in D$, $\sigma_{z_0}(z_0) = 0$. Hence replacing f by $\sigma_\alpha \circ f$, $\alpha = f(0)$, we may suppose that $f(0) = 0$. We shall show that if $f(0) = 0$, then

$$f(z) = (e^{i\theta_1} z_{p(1)}, \ldots, e^{i\theta_n} z_{p(n)}) .$$

By Proposition 2, if $f = (f_1, \ldots, f_n)$, we have

$$f_k(z) = \sum_{j=1}^{n} a_{kj} z_j \ , \quad a_{kj} \in \mathbb{C} .$$

Moreover, if $|z_j| < 1$, then $|f_k(z)| < 1$ (since $f(D) \subset D$). Hence, for $r < 1$, if we choose $z_j = re^{i\psi_j}$ where $a_{kj} = |a_{kj}| e^{-i\psi_j}$, we get

$$\sum_{j=1}^{n} |a_{kj}| < \frac{1}{r} \quad \text{for any } r < 1,$$

i.e.,

$$\text{(I)} \qquad A_k = \sum_{j=1}^{n} |a_{kj}| \leq 1.$$

Consider now, for a given j, the sequence $z_\nu = (0, \ldots, 1 - \frac{1}{\nu}, \ldots, 0)$ where we have $1 - \frac{1}{\nu}$ in the j-th place and 0 elsewhere. By the remark after Definition 3, every limit point of $f(z_\nu)$ in $\mathbb{C}^n$ is on ∂D. Since $f(z_\nu) = (1 - \frac{1}{\nu})(a_{1j}, \ldots, a_{nj}) \to (a_{1j}, \ldots, a_{nj})$, we conclude that this latter point is in ∂D, i.e.,

$$\max_{k=1,\ldots,n} |a_{kj}| = 1 \quad \text{for } j = 1, \ldots, n.$$

Let $k(1)$ be so that $|a_{k(1),1}| = 1$. By (I) above, $a_{k(1),j} = 0$ for $j = 2, \ldots, n$. Let $k(2)$ be so that $|a_{k(2),2}| = 1$. Then $k(2) \neq k(1)$ (since $a_{k(1),2} = 0$) and $a_{k(2),j} = 0$ for $j \neq 2$ by (I). Thus if $k(j)$ is so that $|a_{k(j),j}| = 1$, then $(k(1), \ldots, k(n))$ is a permutation of $(1, \ldots, n)$ and $a_{k(j),i} = 0$ for $i \neq j$. If p is the inverse permutation of the one just constructed, we have $f_k(z) = a_{k,p(k)} z_{p(k)}$, $|a_{k,p(k)}| = 1$. q. e. d.

It is a classical theorem, due to Riemann, that any two simply connected domains in $\mathbb{C}$, $\neq \mathbb{C}$, are analytically equivalent. Proposition 3 enables one to show that the situation is very different for domains in $\mathbb{C}^n$, $n > 1$.

Proposition 4 (Poincaré). Let $P = \{(z_1, z_2) \in \mathbb{C}^2 \mid |z_j| < 1, j = 1, 2\}$ and $B = \{z \in \mathbb{C}^2 \mid |z_1|^2 + |z_2|^2 < 1\}$. Then there is no analytic isomorphism of P onto B.

Proof. Since there is an automorphism of P taking any point of P onto 0, it suffices to show that there is no isomorphism $f: B \to P$ with $f(0) = 0$. Suppose that there is. Then, the map $\sigma \mapsto f \circ \sigma \circ f^{-1}$ is an isomorphism of $\mathrm{Aut}(B)$ onto $\mathrm{Aut}(P)$ as topological groups. Hence, in particular, if G_o denotes the connected component of e of the topological group G, $\mathrm{Aut}(B)_o$ and $\mathrm{Aut}(P)_o$ are isomorphic. Moreover, the above map induces an isomorphism of the subgroup G of $\mathrm{Aut}(B)$ leaving 0 fixed onto the subgroup H of $\mathrm{Aut}(P)$ leaving 0 fixed. Hence G_o is isomorphic to H_o. It follows from Proposition 3 that any $\sigma \in H$ which is sufficiently close to e is of the form

$$(1) \qquad \sigma(z_1, z_2) = (e^{i\theta_1} z_1, e^{i\theta_2} z_2), \quad \theta_1, \theta_2 \in \mathbb{R},$$

(since $\sup |z_2 e^{i\alpha_2} - z_1 e^{i\alpha_1}| \geq 1$ for $|z_1| \leq \frac{1}{2}$, $|z_2| \leq \frac{1}{2}$). Hence H_o contains only elements of the form (1) above, and so is abelian. On the other hand, G_o contains all elements τ of the form $z \to Az$ where A is a 2×2 unitary matrix. Hence the 2×2 unitary group

$U(2) \subset G_o$ as a subgroup. Hence G_o cannot be abelian so that G_o is not isomorphic to H_o.

We shall now prove some theorems, due to Remmert and Stein, which shows that the situation in $n > 1$ variables is really complicated. We begin with a lemma.

Lemma 1. Let Ω be a connected open set in $\mathbb{C}^n$ and $f_1, \dots, f_m$ holomorphic functions on Ω. Suppose that

$$\sum_{j=1}^{m} |f_j(z)|^2 \quad \text{is constant on } \Omega.$$

Then the f_j are all constant.

Proof. If φ is holomorphic on Ω, we have, for $k = 1, \dots, n$,

$$\frac{\partial^2 |\varphi|^2}{\partial z_k \partial \bar{z}_k} = \frac{\partial^2}{\partial z_k \partial \bar{z}_k} \varphi\bar{\varphi} = \frac{\partial}{\partial z_k} (\varphi \frac{\partial \bar{\varphi}}{\partial \bar{z}_k}) = \frac{\partial \varphi}{\partial z_k} \frac{\partial \bar{\varphi}}{\partial \bar{z}_k} = \left| \frac{\partial \varphi}{\partial z_k} \right|^2 .$$

Hence, if $\sum |f_j(z)|^2 = \text{const.}$, we have

$$0 = \frac{\partial^2}{\partial z_k \partial \bar{z}_k} \sum |f_j(z)|^2 = \sum \left| \frac{\partial f_j}{\partial z_k} \right|^2$$

so that $\frac{\partial f_j}{\partial z_k} = 0$ for $j = 1, \dots, m$, $k = 1, \dots, n$.

Theorem 1 (Remmert-Stein). Let D be a domain in $\mathbb{C}^n$, and let $n = n_1 + n_2$, $n_1, n_2 > 0$. Suppose that there is a point $a \in \partial D$ and an open set $U = U_1 \times U_2$, $U_j \subset \mathbb{C}^{n_j}$, containing a such that $U \cap D = D_1 \times D_2$, D_j open and connected in $\mathbb{C}^{n_j}$, with $\bar{D}_2 \cap U_2 \neq U_2$. Then, for any $m \geq 1$, there is no proper holomorphic map of D into the ball $B_m = \{(w_1, \dots, w_m) \in \mathbb{C}^m \mid \sum_{j=1}^{m} |w_j|^2 < 1\}$.

(Note that the condition is, in particular, satisfied if $D = D_1 \times D_2$ and is bounded; we may take $U = \mathbb{C}^n$.)

Proof. Let $f = (f_1, \ldots, f_m)$ be a proper holomorphic map of D into B_m. We shall write $z = (\zeta, \omega)$, $\zeta \in \mathbb{C}^{n_1}$, $\omega \in \mathbb{C}^{n_2}$. Let $\omega_\nu \in D_2$ and suppose that $\omega_\nu \to \omega \in (\partial D_2) \cap U_2$. For $j = 1, \ldots, m$, the functions $\zeta \mapsto f_j(z, \omega_\nu)$ is a holomorphic function $\varphi_{j,\nu}$ on D_1 with $\sum |\varphi_{j,\nu}|^2 < 1$. Hence, by Montel's theorem (chapter 1, Proposition 6), there is a subsequence $\{\nu_k\}$ so that $\varphi_{j,\nu_k} \to \varphi_j$ uniformly on compact subsets of D_1. Now, for any $\zeta \in D_1$, $\{(\zeta, \omega_{\nu_k})\}$ has no limit point in D. Since f is proper, $f(\zeta, \omega_{\nu_k}) = (\varphi_{1,\nu_k}(\zeta), \ldots, \varphi_{m,\nu_k}(\zeta))$ has no limit point in B_m. Since $\sum |\varphi_{j,\nu}(\zeta)|^2 < 1$ on D_1, we conclude that $\sum_{j=1}^{m} |\varphi_j(\zeta)|^2 = \lim_{k\to\infty} \sum_{j=1}^{m} |\varphi_{j,\nu_k}(\zeta)|^2 = 1$ for all $\zeta \in D_1$. Hence by Lemma 1, $\varphi_j = \text{const.}$, $j = 1, \ldots, m$.

Now, by Weierstrass' theorem (chapter 1, Proposition 5), if $\zeta = (\zeta_1, \ldots, \zeta_{n_1})$,

$$\frac{\partial f_j(\zeta, \omega_{\nu k})}{\partial \zeta_p} \to \frac{\partial \varphi_j}{\partial \zeta_p} = 0 \qquad (p = 1, \ldots, n_1).$$

Hence, for $p = 1, \ldots, n_1$, $\dfrac{\partial f_j(\zeta, \omega)}{\partial \zeta_p}$ tends to 0 if ω tends to a point of $(\partial D_2) \cap U_2$. Hence, for fixed $\zeta \in D_1$, the function

$$\omega \to \begin{cases} \partial f_j(\zeta, \omega)/\partial \zeta_p & \text{if } \omega \in D_2 \\ 0 & \text{otherwise} \end{cases}$$

is holomorphic on U_2 (Radó's theorem; chapter 4, Theorem 1). Since, by assumption, $U_2 - \overline{D}_2 \neq \emptyset$, this implies that

$$\frac{\partial f_j}{\partial \zeta_p} \equiv 0 \text{ on } D_1 \times D_2, \quad p = 1, \ldots, n_1 ;$$

hence $\partial f_j / \partial \zeta_p \equiv 0$ on D, $p = 1, \ldots, n_1$. Hence the map f is constant on any connected component of the set $D_1(\omega^o) = \{\omega = \omega^o,\ \omega^o \in D_2\}$. Clearly, $D_1(\omega^o)$ is not relatively compact in D, so that f cannot be proper.

Theorem 2. Let D be a domain in $\mathbb{C}^n$, and let $n = n_1 + n_2$, $n_1, n_2 > 0$. Suppose that there is a point $a \in \partial D$ and an open connected set $U = U_1 \times U_2$, $U_j \subset \mathbb{C}^{n_j}$, containing a such that $U \cap D = D_1 \times D_2$ with D_j open and connected in $\mathbb{C}^{n_j}$ and $\overline{D}_2 \cap U_2 \neq U_2$. Let $W = W_1 \times \ldots \times W_m$ be a domain in $\mathbb{C}^m$ where $W_1, \ldots, W_m$ are bounded domains in $\mathbb{C}$. (We denote a point in $\mathbb{C}^n$ by (z, w), $z \in \mathbb{C}^{n_1}$, $w \in \mathbb{C}^{n_2}$.)

Let $f = (f_1, \ldots, f_m)$ be a proper holomorphic map of D into W. Then, for at least one j, $1 \leq j \leq m$, f_j is locally independent of z, i.e., if $z = (z_1, \ldots, z_{n_1})$, $\partial f_j/\partial z_p = 0$ for $p = 1, \ldots, n_1$.

Proof. We shall need the following generalization of Radó's theorem.

(*) Let $(\varphi_{\mu\nu})$, $1 \leq \mu \leq k$, $1 \leq \nu \leq \ell$ be a matrix of holomorphic functions on $D \subset U$, where D and U are connected open subsets of C^n, and $U - \overline{D} \neq \emptyset$. Suppose that

$$p(z) \equiv \prod_{\nu=1}^{\ell} \sum_{\mu=1}^{k} |\varphi_{\mu\nu}(z)|^2 \to 0 \quad \text{as} \quad z \in D,\ z \to \zeta$$

for any $\zeta \in (\partial D) \cap U$. Then, for some ν_o, $1 \leq \nu_o \leq \ell$, we have

$$\varphi_{\mu\nu_o} \equiv 0 \ , \quad \mu = 1, \ldots k.$$

Proof of (*). Let $a \in (\partial D) \cap U$ and P a polydisc about a contained in U. Let $\alpha \in D \cap P$ and $\beta \in P - \overline{D}$, and let $V \subset \mathbb{C}$ be defined by

$$V = \{\lambda \in \mathbb{C} \mid \alpha + \lambda(\beta-\alpha) \in P\}, \quad V' = \{\lambda \in \mathbb{C} \mid \alpha+\lambda(\beta-\alpha) \in D \cap P\}$$

and let

$$s(\lambda) = p(\alpha + \lambda(\beta-\alpha)) \ , \quad \lambda \in V \ .$$

V is connected. We set

$$u(\lambda) = \begin{cases} \log s(\lambda) \ , & \lambda \in V' \\ -\infty & , \lambda \in V-V' \ . \end{cases}$$

We shall prove below that u is subharmonic in V. It then follows follows from chapter 3, Lemma 2 that $u \equiv -\infty$ on V, i.e., $p(\alpha+\lambda(\beta-\alpha)) = 0$ for $\alpha \in P \cap D$, $\beta \in U - \overline{D}$ and $\lambda \in V$. This clearly implies that $p = 0$ on $D \cap P$, hence on D.

To prove that u is subharmonic on V, it is sufficient, because of chapter 3, Proposition 4, that u is subharmonic on the open set $V' - \{\lambda \in V' \mid u(\lambda) = -\infty\}$, $= V''$ say. On V'' we have

$$\frac{\partial^2 u}{\partial\lambda\partial\overline{\lambda}} = \sum_{\nu=1}^{\ell} \frac{\partial^2}{\partial\lambda\partial\overline{\lambda}} \log \sum_{\mu=1}^{k} |\varphi_{\mu\nu}(\alpha+\lambda(\beta-\alpha))|^2$$

Now, if $f_1, \ldots, f_k$ are holomorphic on V'' and not simultaneously 0, we have

$$\frac{\partial^2}{\partial\lambda\partial\overline{\lambda}} \log \sum_{\mu=1}^{k} |f_\mu(\lambda)|^2$$

$$= \Big(\sum_{\mu=1}^{k} |f_\mu(\lambda)|^2\Big)^{-2}\Big\{\sum_{\mu=1}^{k} |f_\mu(\lambda)|^2 \sum_{\mu=1}^{k} \Big|\frac{df_\mu}{d\lambda}(\lambda)\Big|^2 - \Big|\sum_{\mu=1}^{k} \overline{f}_\mu(\lambda)\frac{df_\mu}{d\lambda}(\lambda)\Big|^2\Big\}$$

which is ≥ 0 by Schwarz's inequality. This proves our assertion, and with it, (*).

Proof of Theorem 2. Let $\{w_\nu\}$ be a sequence of points of D_2 converging to a point $w^0 \in (\partial D_2) \cap U_2$. Then, there is a subsequence $\{\nu_k\}$ so that if $\varphi_{j,\nu}(z) = f_j(z, w_\nu)$, then $\{\varphi_{j,\nu_k}\}$ converges uniformly on compact subsets of D_1 to a function φ_j (Montel's theorem; the W_j

are bounded). Since f is proper, no limit point of $f(z, w_\nu)$ can be in W. Hence, for all $z \in D_1$, $(\varphi_1(z), \ldots, \varphi_m(z)) \in \partial W$. Set

$$E_j = \{z \in D_1 \mid \varphi_j(z) \in \partial W_j\}.$$

Then $\bigcup E_j = D_1$. Since each E_j is obviously closed in D_1, it follows that E_j has a nonempty interior V for some j, $1 \le j \le m$. Since φ_j maps V into ∂W_j and a nonconstant holomorphic function is an open map into $\mathbb{C}$, it follows that φ_j is constant. [Note that the index j depends, a priori, on the sequences $\{w_\nu\}, \{w_{\nu_k}\}$.] Hence $\frac{\partial \varphi_j}{\partial z_\nu} \equiv 0$ on D_1, $\nu = 1, \ldots, n_1$ $(z = (z_1, \ldots, z_{n_1}))$. In any case, it follows that for any sequence $\{w_\nu\}$ tending to a point of $(\partial D_2) \cap U_2$, a subsequence of

$$\prod_{j=1}^{m} \sum_{\nu=1}^{n_1} \left| \frac{\partial f_j}{\partial z_\nu} (z, w_\nu) \right|^2$$

tends to zero. Hence

$$\prod_{j=1}^{n} \sum_{\nu=1}^{n_1} \left| \frac{\partial f_j}{\partial z_\nu} (z, w) \right|^2$$

tends to 0 as w tends to a point of $(\partial D_2) \cap U_2$. It follows from (*) that we have:

(**) For each $z \in D_1$, there is $j = j(z)$ such that

$$p_j(z, w) = \sum_{\nu=1}^{n_1} \left| \frac{\partial f_j}{\partial z_\nu} (z, w) \right|^2 = 0 \quad \text{for all } w \in D_2 .$$

Since the set $F_j = \{(z, w) \in D_1 \times D_2 \mid p_j(z, w) = 0\}$ is closed and $\bigcup F_j = D_1 \times D_2$, at least one F_j has an interior point. Clearly

$$\frac{\partial f_j}{\partial z_\nu}(z, w) \equiv 0 \ \text{ on } D \ , \quad \nu = 1, \ldots, n_1 \ ,$$

for this value of j (principle of analytic continuation).

Lemma 2. Let Ω be an open set in $\mathbb{C}^n$, $n \geq 2$. Then there is no proper holomorphic map of Ω into a domain $W \subset \mathbb{C}$.

Proof. Let $a \in \Omega$ and $A = f^{-1}f(a)$. Then A is compact. Let $g(z) = (f(z) - f(a))^{-1}$; g is holomorphic on $\Omega - A$. Let B be the closed ball about 0 with the smallest radius containing A (in general, $B \not\subset \Omega$). Then, there exists $z_o \in (\partial B) \cap A$. We may suppose that B has radius 1. By choosing a suitable orthonormal basis of $\mathbb{C}^n$, we may suppose that $z_o = (0, \ldots, 0, 1)$. Then if $\delta > 0$ is sufficiently small, the set $|z_1| = \delta$, $z_2 = 0$, $z_{n-1} = 0$, $z_n = 1$ is contained in $\Omega - A$. Hence, if $\varepsilon > 0$ is small enough, so is the set $\delta - \varepsilon \leq |z_1| \leq \delta + \varepsilon$, $|z_j| \leq \varepsilon$, $j = 2, \ldots, n-1$, $|z_n - 1| \leq \varepsilon$, so that g is holomorphic on this set. Moreover, g is holomorphic on the set $|z_1| \leq \delta + \varepsilon$, $|z_j| \leq \varepsilon$, $j = 2, \ldots, n-1$, $z_n = 1 + \eta$, $\eta > 0$ sufficiently small (since these points have a distance $\geq |z_n| = 1 + \eta > 1$ from 0 and B has radius 1). Hence g can be continued holomorphically to a neighborhood of z_o (by chapter 4, Theorem 2) which is absurd since $\Omega - A$ is dense and $g \to \infty$ as $z \to z_o$.

Theorem 3. Let $D = D_1 \times D_2 \subset \mathbb{C}^2$, $D_j \subset \mathbb{C}$ and $W = W_1 \times W_2$, $W_j \subset \mathbb{C}$ be bounded open sets in $\mathbb{C}^2$. Let $f = (f_1, f_2)$ be a proper holomorphic map of D into W. Then f_j is of the form $f_j(z_{p(j)})$, $j = 1, 2$ where p is a permutation of $(1, 2)$.

Proof. By Theorem 2, if $z = (z_1, z_2) \in \mathbb{C}^2$, one of the f_j is independent of z_1, one of z_2. It is sufficient to prove that neither f_j is independent of z_1 and of z_2. But if, for instance, f_1 is constant, $f_2 : D \to W_2$ would be a proper holomorphic map, contradicting Lemma 2.

Theorem 3 applies in particular to analytic isomorphisms of D onto W. The most general result that follows easily from the reasoning given in Theorem 2 is the following.

Proposition I. Let D_j, W_j be bounded domains in $\mathbb{C}^{n_j}$, $j = 1, \ldots, k$, and suppose the following about W_j:

∂W_j contains no positive dimensional analytic set in an open set in $\mathbb{C}^{n_j}$.

Let $f = (f_1, \ldots, f_k)$ be a proper holomorphic map of $D = D_1 \times \ldots \times D_k$ into $W_1 \times \ldots \times W_k$; here f_j is a map of D into $\mathbb{C}^{n_j}$. Denote a point $z \in D$ by $(z_1, \ldots, z_k)$, $z_j \in \mathbb{C}^{n_j}$. Then, there is a permutation p of the set $\{1, \ldots, k\}$ and proper maps

$$g_j : D_j \to W_{p(j)}$$

such that

$$f = g_1 \times \ldots \times g_k .$$

The proof uses the following generalization of Lemma 2.

If $\Omega \subset \mathbb{C}^n$ and $W \subset \mathbb{C}^m$, $m < n$, are open sets, there is no proper holomorphic map of Ω into W.

For a proof of this latter fact, see [20], chapter III, Proposition 10 and its corollaries, and chapter VII, Propositions 1 and 2.

Further, in the case of automorphisms, H. Cartan has proved the following generalization of the proposition above (see [8]).

Proposition II (H. Cartan). Let $D = D_1 \times D_2$, $D_j \subset \mathbb{C}^{n_j}$, $n = n_1 + n_2$, $n_1, n_2 > 0$, be a bounded domain. Then, any $f \in \mathrm{Aut}\,(D)$ which belongs to the connected component of the identity in Aut (D) is of the form

$$f = f_1 \times f_2 \qquad \text{where } f_j \in \text{Aut}(D_j).$$

We shall prove now a fundamental theorem, due again to H. Cartan, and give some of its applications.

Theorem 4 (H. Cartan). Let D be a bounded domain in $\mathbb{C}^n$. Let $\{f_\nu\} \subset \text{Aut}(D)$ be a sequence of automorphisms of D. Suppose that f_ν converges, uniformly on compact subsets of D, to a holomorphic map $f: D \to \mathbb{C}^n$. Then, the following three properties are equivalent.

(i) $f \in \text{Aut}(D)$

(ii) $f(D) \not\subset \partial D$

(iii) There exists $a \in D$ such that the jacobian matrix $(\frac{\partial f_i}{\partial z_j}(a))$, $f = (f_1, \ldots, f_n)$, has a nonzero determinant.

We shall need two preliminary propositions.

If $f: \Omega \to \mathbb{C}^m$ is a holomorphic map, we denote by $(df)_a$, $a \in \Omega$, the linear map of $\mathbb{C}^n$ into $\mathbb{C}^m$ defined by

$$(df)_a(v_1, \ldots, v_n) = (w_1, \ldots, w_m),$$

where

$$w_j = \sum_{k=1}^{n} v_k \frac{\partial f_j}{\partial z_k}(a) \ , \quad f = (f_1, \ldots, f_m).$$

Lemma 3. Let $\Omega \subset \mathbb{C}^n$ and $f: \Omega \to \mathbb{C}^n$ be a holomorphic map. Suppose that $\det(df)_a \neq 0$ for some $a \in \Omega$. Then there exists neighborhoods U of a and V of $f(a)$ such that $f(U) \subset V$ and $f|U$ is an analytic isomorphism onto V.

This is a special case of the implicit function theorem which we do not prove. For a proof, see, for example, [21, chapter 1].

Proposition 5. Let $\{\varphi_\nu\}$ be a sequence of continuous open mappings of $\Omega \subset \mathbb{C}^n$ into $\mathbb{C}^n$. Suppose that φ_ν converges, uniformly on compact subsets of Ω to a map $\varphi:\Omega \to \mathbb{C}^n$.

Suppose that, for some $a \in \Omega$, a is an isolated point of $\varphi^{-1}\varphi(a)$. Then, for any neighborhood U of a, there is a ν_o such that $\varphi(a) \in \varphi_\nu(U)$ for $\nu \geq \nu_o$.

Proof. Suppose that this is false. By passing to a subsequence, we may suppose that U is such that

$$\overline{U} \text{ is compact, } \varphi(a) \notin \varphi_\nu(\overline{U}), \quad \nu = 1, 2, \ldots,$$

$$\overline{U} \cap \varphi^{-1}\varphi(a) = \{a\}.$$

Then $\varphi(\partial U)$ is compact and $\varphi(a) \notin \varphi(\partial U)$. Hence there is a neighborhood V of $\varphi(\partial U)$ and a polydisc P about $\varphi(a)$ so that

$$(*) \qquad P \cap V = \emptyset.$$

Now, if ν_o is large enough, $\varphi_\nu(\partial U) \subset V$ for $\nu \geq \nu_o$. Since φ_ν is open, we have $\partial\varphi_\nu(U) \subset \varphi_\nu(\partial U)$. Hence $\varphi_\nu(U)$ is a relatively compact open set in $\mathbb{C}^n$ with $\partial\varphi_\nu(U) \subset V$. We claim now that $\{\partial\varphi_\nu(U)\} \cap P \neq \emptyset$ if ν is large enough (which would contradict (*) above and so end the proof). Since $\varphi(a) \in P$ and $\varphi_\nu(a) \to \varphi(a)$, if ν is large, $\varphi_\nu(a) \in P$. On the other hand, $\varphi(a) \notin \varphi_\nu(U)$ by assumption. If $\{\partial\varphi_\nu(U)\} \cap P = \emptyset$, we would have

$$P = \{\varphi_\nu(U) \cap P\} \cup \{(\mathbb{C}^n - \overline{\varphi_\nu(U)}) \cap P\}$$

and each of the open sets above is nonempty [$\varphi_\nu(a)$ belongs to the first, $\varphi(a)$ to the second] contradicting the fact that P is connected. Thus $\{\partial\varphi_\nu(U)\} \cap P \neq \emptyset$ and the proposition is proved.

Corollary (Hurwitz's theorem). Let Ω be an open connected set in $\mathbb{C}^n$ and $\{f_\nu\}$ a sequence of holomorphic functions on Ω, converging uniformly on compact sets to a holomorphic function f. Then if $f_\nu(z) \neq 0$ for all ν, and all z, and f is nonconstant, we have $f(z) \neq 0$ for all $z \in \Omega$.

Proof. Suppose that $f(a) = 0$, $a \in \Omega$. Let P be a small polydisc about a. Then $f \not\equiv 0$ on P (since if it were, f would be 0 on Ω since Ω is connected). Let $b \in P$, $f(b) \neq 0$. Let $D = \{\lambda \in \mathbb{C} \mid a + \lambda(b-a) \in P\}$. Then D is a convex, hence connected open set in $\mathbb{C}$. Let $\varphi_\nu(\lambda) = f_\nu(a + \lambda(b-a))$, $\varphi(\lambda) = f(a + \lambda(b-a))$. Then $\varphi(0) = 0$, $\varphi(1) = f(b) \neq 0$, so that φ is nonconstant on D. Hence for large ν, φ_ν is also nonconstant, hence an open map of D into $\mathbb{C}$ (by chapter 1, Proposition 4). By Proposition 5 above, $f_\nu(\Omega) \supset \varphi_\nu(D) \supset \{0\}$ if ν is large, a contradiction.

Proof of Theorem 4. (i) $\Rightarrow$ (ii). Obvious.

(i) $\Rightarrow$ (iii). If $f \in \mathrm{Aut}(D)$ and $a \in D$, and $g = f^{-1} \in \mathrm{Aut}(D)$, we have

$$g \circ f = \text{identity},$$

hence $(dg)_{f(a)} \circ (df)_a = \text{identity}$, so that $(df)_a$ is invertible.

(iii) $\Rightarrow$ (ii). If $(df)_a$ has a nonzero determinant, then, by Lemma 3, f(D) contains a (nonempty) neighborhood of $f(a)$, hence $f(D) \not\subset \partial D$.

(ii) $\Rightarrow$ (iii). Clearly $f(D) \subset \overline{D}$. Hence, if (ii) holds, $f(D) \cap D \neq \emptyset$. Let $a \in D$ be so that $f(a) = b \in D$. Let $g_\nu = f_\nu^{-1}$, and let $\{\nu_k\}$ be a subsequence so that $\{g_{\nu_k}\}$ converges uniformly on compact subsets of D to $g: D \to \mathbb{C}^n$ (Montel's theorem, chapter 1, Proposition 6).

We have

$$g(b) = \lim_{k\to\infty} f_{\nu_k}^{-1}(f(a)) .$$

Moreover, if k is large, $f_{\nu_k}(a)$ is close to $f(a)$, hence in a compact subset of D. Since $f_{\nu_k}^{-1}$ converges uniformly on compact subsets of D, we deduce that

$$g(b) = \lim_{k\to\infty} f_{\nu_k}^{-1}(f_{\nu_k}(a)) = \lim_{k\to\infty} a = a.$$

Thus $g(b) = a \in D$. Let V be a small neighborhood of b. Then $g(V)$ lies in a compact subset of D, hence, there is K compact in D so that $g_{\nu_k}(V) \subset K$ (k large). Then, for $x \in V$, we have

$$\begin{aligned} f(g(z)) = \lim_{k\to\infty} f(g_{\nu_k}(x)) &= \lim_{k\to\infty} f_{\nu_k}(g_{\nu_k}(x)) \quad (\text{since } g_{\nu_k}(V) \subset K \\ &\qquad\qquad \text{and } f_{\nu_k} \to f \text{ uniformly on } K) \\ &= x . \end{aligned}$$

Hence $(df)_{g(x)} \circ (dg)_x$ = identity for $x \in V$; in particular, $\det((df)_y) \neq 0$ for $y \in g(V)$, which proves (iii).

It remains to prove that

(iii) $\Rightarrow$ (i). The function $j_\nu(x) = \det(df_\nu)_x$ is holomorphic on D and converges to $j(x) = \det(df)_x$, uniformly on compact subsets of D. Moreover, if (iii) holds, $j(x) \neq 0$. Also $j_\nu(x) \neq 0$ for all ν, and all x since $f_\nu \in \text{Aut}(D)$ (see proof that (i) $\Rightarrow$ (iii)). If $j(x)$ is constant, $j(x)$ is obviously never 0, and if $j(x)$ is nonconstant, it is again never 0 by the corollary to Proposition 5 above. Hence, in either case, $j(x) \neq 0$ for all $x \in D$. By Lemma 3, $f: D \to \mathbb{C}^n$ is an open map and any $x \in D$ is isolated in $f^{-1}f(x)$. It follows, from Proposition 5, that $f(D) \subset \bigcup f_\nu(D) = D$.

Let $\{\nu_k\}$ be a subsequence of $\{\nu\}$ so that g_{ν_k} converges uniformly on compact subsets of D. Then, for $x \in D$, $\{f_{\nu_k}(x)\}$ converges to $f(x) \in D$, hence lies in a compact subset of D. Hence

$$g(f(x)) = \lim_{k\to\infty} g_{\nu_k}(f_{\nu_k}(x)) = x, \quad \text{for all } x \in D.$$

In particular, $\det(dg)_y \neq 0$ for $y \in f(D)$. Hence, repeating our argument above, we conclude that $g(D) \subset D$, so that $g_{\nu_k}(x)$ lies in a compact subset of D for any $x \in D$. Hence

$$f(g(x)) = \lim_{k\to\infty} f_{\nu_k}(g_{\nu_k}(x)) = x\ .$$

Thus $f \circ g = \text{identity}$, $g \circ f = \text{identity}$, and we conclude that $f \in \text{Aut}(D)$.

We shall now give some applications of this theorem.

Proposition 6. Let D be a bounded domain in $\mathbb{C}^n$ and K, L compact sets in D. Then the set

$$G(K, L) = \{f \in \text{Aut}(D) \mid f(K) \cap L \neq \emptyset\}$$

is compact.

Proof. Let $\{f_\nu\}$ be a sequence of elements of G(K, L). By passing to a subsequence, we may suppose that f_ν converges to a map of D into $\mathbb{C}^n$ (Montel's theorem). Since $f_\nu(K) \cap L \neq \emptyset$, there is $a_\nu \in K$ so that $f(a_\nu) = b_\nu \in L$. If $\{\nu_k\}$ is a subsequence so that $a_{\nu_k} \to a \in K$, $b_{\nu_k} \to b \in L$ (K, L are compact), then $f(a) = b$, so that $f \in \text{Aut}(D)$ by Theorem 4 and since $f(a) = b$, $f \in G(K, L)$. Since any sequence of elements in G(K, L) contains a subsequence which converges in G(K, L), this set is compact.

Proposition 7. If D is a bounded domain, $\mathrm{Aut}(D)$ is a locally compact group.

Proof. If K, L are compact sets in D with $K \subset \overset{\circ}{L}$, then $G(K, L)$ is a neighborhood of the identity (by definition of the topology on $\mathrm{Aut}(D)$) which is compact by Proposition 6.

Definition 4. Let G be a topological group and X a (Hausdorff) topological space. We say that G operates on X if we are given a continuous map $G \times X \to X$, $(g, x) \mapsto g \cdot x$ such that $ex = x$ for all $x \in X$ and $(gg')x = g(g'x)$ for all $g, g' \in G$, $x \in X$.

If G and X are locally compact, we say that G acts properly on X if the map $G \times X \to X \times X$ defined by $(g, x) \mapsto (gx, x)$ is proper.

If G is discrete and X is locally compact, we say that G acts properly discontinuously on X if, for any $a \in X$, there is a neighborhood U of a so that $\{g \in G \mid g(U) \cap U \neq \emptyset\}$ is finite.

Remark. If G, X are locally compact, G acts properly on X if and only if, for any compact sets $K, L \subset X$, the set $G(K, L) = \{g \in G \mid g(K) \cap L \neq \emptyset\}$ is compact.

In fact, suppose this condition is satisfied. Any compact set in $X \times X$ is contained in a set of the form $K \times K$, $K \subset X$ compact. The inverse image of $K \times K$ by the map $(g, x) \mapsto (gx, x)$ is just $G(K, K)$ and so is compact.

Conversely, if the map $(g, x) \mapsto (gx, x)$ is proper, then, as above, $G(K, K)$ is compact and $G(K, L) \subset G(A, A)$, $A = K \cup L$.

A discrete group G acts properly discontinuously if and only if $G(K, L)$ is finite for any compact $K, L \subset X$.

Proposition 8. If D is a bounded domain in $\mathbb{C}^n$, $\mathrm{Aut}(D)$ acts properly on D.

This follows at once by the remark above and Proposition 6.

Proposition 9. A subgroup $\Gamma \subset \mathrm{Aut}(D)$, provided with the discrete topology, acts properly discontinuously on D if and only Γ is a discrete subgroup of $\mathrm{Aut}(D)$ [i.e., a discrete subset].

Proof. If Γ is a discrete subgroup of $\mathrm{Aut}(D)$ and K, L compact in D, then $\Gamma(K, L) = \{\gamma \in \Gamma \mid \gamma(K) \cap L \neq \emptyset\}$ is relatively compact in $\mathrm{Aut}(D)$ by Proposition 6; since Γ is discrete, hence closed, it is a compact, hence finite, subset of Γ.

Conversely, if Γ acts properly discontinuously and K, L are compact in D, $K \subset \overset{\circ}{L}$, then $\Gamma(K, L)$ is a neighborhood of the unit element of Γ (since it contains the projection on Γ of the inverse image by the map $(\gamma, x) \longmapsto (\gamma x, x)$ of the open set $\overset{\circ}{L} \times U$, where U is open in D and $K \subset U \subset \overline{U} \subset \overset{\circ}{L}$). Moreover, $\Gamma(K, L)$ is finite (since Γ acts properly discontinuously). Hence Γ is a discrete subgroup of $\mathrm{Aut}(D)$.

Proposition 10. Let D be a bounded domain in $\mathbb{C}^n$ and $\Gamma \subset \mathrm{Aut}(D)$ a discrete subgroup. Let D/Γ be the quotient of D by the equivalence relation: $x \sim y$ if there exists $\gamma \in \Gamma$ such that $\gamma x = y$.

If D/Γ is compact, Γ is finitely generated.

Proof. Let $\{U_\nu\}$ be a sequence of relatively compact open sets in D such that $\overline{U}_\nu \subset U_{\nu+1}$, $\bigcup U_\nu = D$. Let $\pi: D \to D/\Gamma$ denote the natural projection. Then π is open, so that $V_\nu = \pi(U_\nu)$ is open in D/Γ and $\bigcup V_\nu = D/\Gamma$. Since $V_\nu \subset V_{\nu+1}$, and D/Γ is compact, there is p so that $V_p = D/\Gamma$. This implies that $D = \bigcup_{\gamma \in \Gamma} \gamma(K)$, where $K = \overline{U}_p$ is compact in D.

Let $\{\gamma_1, \ldots, \gamma_N\}$ be the elements of Γ for which $\gamma(K) \cap K \neq \emptyset$. Clearly, γ_i^{-1} is a γ_j, $i = 1, \ldots, N$. We claim that any $\gamma \in \Gamma$ can be written in the form $\gamma = \gamma_{i_1} \cdots \gamma_{i_p}$, $1 \leq i_k \leq N$.

Let Γ' be the subgroup generated by $\{\gamma_1, \ldots, \gamma_N\}$. Since $\gamma_i^{-1} \in \{\gamma_1, \ldots, \gamma_N\}$ for $1 \leq i \leq N$, Γ' is the set of products $\gamma_{i_1} \cdots \gamma_{i_p}$. If $\Gamma' \neq \Gamma$, let $\Gamma'' = \Gamma - \Gamma'$. Let $\Gamma'(K) = \bigcup_{\gamma' \in \Gamma'} \gamma'(K)$, $\Gamma''(K) = \bigcup_{\gamma'' \in \Gamma''} \gamma''(K)$. Then, since $\Gamma = \Gamma' \cup \Gamma''$, and $\bigcup_{\gamma \in \Gamma} \gamma(K) = D$, we have $\Gamma'(K) \cup \Gamma''(K) = D$. Further, $\Gamma'(K) \cap \Gamma''(K) = \emptyset$. In fact, if $\gamma' x = \gamma'' y$, $x, y \in K$, $\gamma' \in \Gamma'$, $\gamma'' \in \Gamma''$, we would have $\gamma y = x$, $\gamma = \gamma'^{-1}\gamma''$. Since $x, y \in K$, $\gamma \in \{\gamma_1, \ldots, \gamma_N\}$, say $\gamma = \gamma_i$. Then $\gamma'' = \gamma'\gamma_i \in \Gamma'$, a contradiction. Hence $\Gamma'(K), \Gamma''(K)$ are disjoint.

Moreover, the family $\{\gamma(K)\}_{\gamma \in \Gamma}$ is locally finite (i.e., any point of D has a neighborhood U such that $\{\gamma \in \Gamma \mid \gamma(K) \cap U \neq \emptyset\}$ is finite; it suffices to take for U any compact neighborhood. Hence $\Gamma'(K)$ and $\Gamma''(K)$ are closed in D. Since D is connected and $\Gamma'(K) \neq \emptyset$, this implies that $\Gamma''(K) = \emptyset$, i.e., $\Gamma'' = \emptyset$, so that $\Gamma' = \Gamma$ and the result is proved.

We proceed now to prove the theorem of Osgood referred to at the beginning of this chapter. We shall need the rank theorem (which is stronger than Lemma 3). For a proof, see e.g., [21, chapter 1].

<u>The rank theorem.</u> Let Ω be an open set in $\mathbb{C}^n$ and $f:\Omega \to \mathbb{C}^m$ a holomorphic map. Suppose that the rank of the linear map $(df)_a$ is an integer $k \leq n$ independent of $a \in \Omega$. Then, for any $a \in \Omega$, there exist neighborhoods U of a, V of $f(a)$, polydiscs P about 0 in $\mathbb{C}^n$, Q about 0 in $\mathbb{C}^m$ respectively, and analytic isomorphisms $u: P \to U$ and $v: V \to Q$ so that the map $v \circ f \circ u: P \to Q$ is given by $(z_1, \ldots, z_n) \longrightarrow (z_1, \ldots z_k, 0, \ldots, 0)$. In particular, if $k < n$, no point $a \in \Omega$ is isolated in $f^{-1}f(a)$.

<u>Theorem 5.</u> Let Ω be an open set in $\mathbb{C}^n$ and $f:\Omega \to \mathbb{C}^n$ an injective holomorphic map. Then f is a homeomorphism of Ω onto an open set $\Omega' \subset \mathbb{C}^n$ and the inverse map $f^{-1}:\Omega' \to \Omega$ is holomorphic.

<u>Proof.</u> We may suppose that Ω is connected. We first assert that there exists $a \in \Omega$ such that $(df)_a$ has rank n (so that $\det((df)_a) \neq 0$). Suppose that this is false, and let $k = \max_{a \in \Omega} \operatorname{rank}(df)_a < n$. Let $z_0 \in \Omega$ be such that rank $(df)_{x_0} = k$. Then clearly, there is a neighborhood U of x_0 such that $\operatorname{rank}(df)_x \geq k$ (hence $= k$ since k is the maximum rank) for $x \in U$. By the rank theorem, x_0 cannot be isolated in $f^{-1}f(x_0)$, hence there is $x_1 \in U$ with $f(x_1) = f(x_0)$, $x_0 \neq x_1$, contradicting our assumption that f is injective.

Let $A = \{x \in \Omega \mid \det(df)_x = 0\}$. Since $x \longmapsto \det(df)_x$ is obviously holomorphic, and, by what we have seen above, $A \neq \Omega$, it follows that

A is an analytic set in Ω, and $\Omega - A$ is dense in Ω. If we prove that $A = \emptyset$, the theorem follows from Lemma 3.

Let $a \in \Omega$, and let P be a small polydisc about a, $\overline{P} \subset \Omega$. Then ∂P is compact, hence so is $f(\partial P) = K$. Since $a \notin \partial P$ and f is injective, $f(a) \notin K$. Let V be a small polydisc about $f(a)$, $\overline{V} \cap K = \emptyset$, and let $U = f^{-1}(V) \cap P$. We assert that the map $f: U \to V$ is proper. In fact, if C is compact in V, $f^{-1}(C)$ cannot be adherent to ∂P (since there is a neighborhood N of ∂P with $f(N) \cap V = \emptyset$), hence is relatively compact in P. It follows that $f^{-1}(C) \cap U$ is compact.

Let $W = f(U - A)$. By Lemma 3, $f: U - A \to W$ is an analytic isomorphism (and W is open in $\mathbb{C}^n$). Let $g: W \to U - A$ be the inverse of $f|U - A$. Let $\varphi(z) = \det((df)_z)$, $z \in U$. φ is holomorphic on U, and $U \cap A = \{z \in U \mid \varphi(z) = 0\}$. Let $\psi(x) = \det((dg)_x)$, $x \in W$. Then ψ is holomorphic on W, and, since $f \circ g =$ identity on W, we have

$$\varphi(g(x)) \cdot \psi(x) = 1 \ , \quad x \in W \ .$$

Let $a \in (\partial W) \cap V$ and $x_\nu \in W$, $x_\nu \to a$, $\nu \to \infty$. Then $g(x_\nu) = f^{-1}(x_\nu)$ is in a compact subset of U (since $f: U \to V$ is proper). Further, every limit point of $g(x_\nu)$ lies on $A \cap U$. Hence, since φ is holomorphic on U and $\varphi | A \cap U = 0$, it follows that $\varphi(g(x_\nu)) \to 0$ as $\nu \to \infty$. Hence, the function $1/\psi$ on W tends to 0 as we tend to a point of $(\partial W) \cap V$. By Radó's theorem (chapter 4, Theorem 1), the function

$$\eta(x) = \begin{cases} 1/\psi(x) & , \ x \in W \ , \\ 0 & , \ x \in V - W \end{cases}$$

is holomorphic on V and $B = V - W = \{x \in V \mid \eta(x) = 0\}$ is an analytic subset of V, $B \neq V$. Hence W is dense in V. Further, if $g = (g_1, \ldots, g_n)$, the g_j are bounded on W since $U \subset P$ is bounded.

Hence, by chapter 4, Proposition 2 (Riemann's continuation theorem), there is a holomorphic map $G: V \to \overline{U} \subset \Omega$ such that $G|W = g$. Then $f \circ G$ is the identity on W since $f \circ G|W = f \circ g =$ identity (and W is dense). This proves that $(f|U)^{-1} = G$ is holomorphic, and the theorem follows

The basic theorems of Cartan are in [5] (Propositions 1 and 2). The theorem on limits of automorphisms in [6] and the result, stated without proof in this chapter, on the automorphisms of a product will be found in [8].

For Theorem 5 about the inverse of an injective holomorphic map of $\Omega \subset \mathbb{C}^n$ into $\mathbb{C}^n$, see Osgood [23] and the references given there.

The results of Remmert-Stein will be found in [28]. They prove Theorem 2 (see also Proposition I) only in the case of products of two plane domains, since Radó's theorem applies directly only to this case.

Several references to related results will be found in [1].

Chapter 6

ANALYTIC CONTINUATION: ENVELOPES OF HOLOMORPHY

We have seen, in chapter 2, that the domain of existence of a holomorphic function in a domain in $\mathbb{C}^n$ can be constructed as a domain over $\mathbb{C}^n$. We shall now develop analogues of these results for a family of holomorphic functions.

Let S be a set. In analogy with the considerations of chapter 2, we define the sheaf $\mathcal{O}(S)$ of S-germs of holomorphic functions on $\mathbb{C}^n$ as follows.

Let U be an open set in $\mathbb{C}^n$ and $\{f_s\}_{s \in S}$ a family of holomorphic functions on U, indexed by S. For $a \in \mathbb{C}^n$ and $(U, \{f_s\})$, $(V, \{g_s\})$ with $a \in U$, $a \in V$, we say that these two pairs are equivalent if there exists a neighborhood W of a, $W \subset U \subset V$ such that, for all $s \in S$, $f_s | W = g_s | W$. An equivalence class with respect to this relation is called an S-germ of holomorphic functions at a. We denote the set of S-germs at a by $\mathcal{O}_a(S)$. We set $\mathcal{O}(S) = \bigcup_{a \in \mathbb{C}^n} \mathcal{O}_a(S)$. We have a natural projection $p = p_S: \mathcal{O}(S) \to \mathbb{C}^n$ defined by $p(g) = a$ if $g \in \mathcal{O}_a(S)$. We define a topology on $\mathcal{O}(S)$ as follows (see chapter 2 for the case when S reduces to a single element). Let $\underline{g}_a \in \mathcal{O}_a(S)$ and let $(U, \{f_s\})$ be a representant of $\underline{g}_a$. Let $\underline{g}_b$ be the S-germ defined by $\{f_s\}$ at $b \in U$, and let $N(U, \{f_s\}) = \bigcup_{b \in U} \underline{g}_b$. The sets $N(U, \{f_s\})$ form a fundamental system of neighborhoods of $\underline{g}_a$. We prove, as in chapter 2, the following:

Proposition 1. The map $p: \mathcal{O}(S) \to \mathbb{C}^n$ is continuous and is a local homeomorphism of $\mathcal{O}(S)$ onto $\mathbb{C}^n$. Further, $\mathcal{O}(S)$ is a Hausdorff topological space. The triple $(\mathcal{O}(S), p, \mathbb{C}^n)$ is an unramified domain over $\mathbb{C}^n$.

Let $p: X \to \mathbb{C}^n$, $p': X' \to \mathbb{C}^n$ be domains over $\mathbb{C}^n$. A continuous map $u: X \to X'$ is called a local isomorphism (or local analytic iso morphism) if every point $a \in X$ has a neighborhood U such that $u|U$ is a homeomorphism onto an open set $U' \subset X'$ and $u|U$ and $(u|U)^{-1}$ are holomorphic (on U and U' respectively). If, in addition, u is a homeomorphism of X onto X', we say that u is an isomorphism (or analytic isomorphism).

If $u: X \to X'$ is a continuous map such that $p' \circ u = p$, then u is automatically a local isomorphism.

Note that the analogue of chapter 5, Lemma 3 holds for holomorphic maps $\Omega \to \Omega'$.

Definition 1. Let $p_o: \Omega \to \mathbb{C}^n$ be a connected domain and $S \subset \mathcal{H}(\Omega)$. Let $p: X \to \mathbb{C}^n$ be a connected domain and $\varphi: \Omega \to X$ a continuous map with $p \circ \varphi = p_o$. We say that $p: X \to \mathbb{C}^n$, $\varphi: \Omega \to X$ is an S-extension of $p_o: \Omega \to \mathbb{C}^n$ if, to every $f \in S$, there is $F_f \in \mathcal{H}(X)$ such that $F_f \circ \varphi = f$.

Note that F_f is uniquely determined (first on $\varphi(\Omega)$ since $F_f \circ \varphi = f$, hence on X by analytic continuation). It is called the extension, (or continuation) of f to X.

Definition 2. Let $p_o:\Omega \to \mathbb{C}^n$ be a connected domain over $\mathbb{C}^n$ and $S \subset \mathcal{H}(X)$. An S-envelope of holomorphy is an S-extension $p:X \to \mathbb{C}^n$, $\varphi:\Omega \to X$ such that the following holds:

(*) For any S-extension $p':X' \to \mathbb{C}^n$, $\varphi':\Omega \to X'$ of $p_o:\Omega \to \mathbb{C}^n$, there is a holomorphic map $u: X' \to X$ such that $p' = p \circ u$, $\varphi = u \circ \varphi'$ and $F_f' = F_f \circ u$, for all $f \in S$, where F_f, F_f' are the extensions of $f \in S$ to X, X' respectively.

Note that u in (*) is unique (since it is determined on $\varphi'(\Omega)$ by the equation $u \circ \varphi' = \varphi$).

It is sufficient to require that $u \circ \varphi' = \varphi$ at one point; equality on Ω is a consequence. If the functions $\{F_f\}_{f \in S}$ separate the points of $p^{-1}p(a)$ for a point $a \in \varphi(\Omega)$, the equation $\varphi = u \circ \varphi'$ is a consequence of the other requirements of (*).

Remark. The S-envelope of holomorphy, if it exists, is unique up to "isomorphism". In fact, let $p: X \to \mathbb{C}^n$, $\varphi:\Omega \to X$ and $p':X' \to \mathbb{C}^n$, $\varphi':\Omega \to X'$ be two S-envelopes of holomorphy. Then, by (*) of Definition 2, there are holomorphic maps $u: X' \to X$, $v: X \to X'$ such that $p = p' \circ v$, $p' = p \circ u$, $\varphi = u \circ \varphi'$, $\varphi' = v \circ \varphi$. Then $u \circ v \circ \varphi = u \circ \varphi' = \varphi$, so that $u \circ v$ is the identity on $\varphi(\Omega)$ which is open in X. Hence, by chapter 2, Proposition 5, $u \circ v$ = identity on X. Similarly, $v \circ u$ = identity on X'. Thus, u is an isomorphism of X' onto X with $p' = p \circ u$, $\varphi = u \circ \varphi'$, which is the uniqueness.

Theorem 1 (Thullen). The S-envelope of holomorphy of any $S \subset \mathcal{H}(\Omega)$ exists.

Proof. For any $p_0 : \Omega \to \mathbb{C}^n$ and $S \subset \mathcal{H}(\Omega)$, we define a map $\varphi = \varphi(p_0, S)$ of Ω into $\mathcal{O}(S)$ as follows. Let $a \in \Omega$ and $a_0 = p_0(a) \in \mathbb{C}^n$. Let U be an open neighborhood of a such that $p_0 | U$ is an isomorphism onto an open set $U_0 \subset \mathbb{C}^n$. Let $\underline{g}_a$ be the S-germ at a_0 defined by the pair $(U_0, \{f_s\})$, where $f_s = s \circ (p_0 | U)^{-1}$, $s \in S$. We set $\varphi(a) = \underline{g}_{a_0}$.

One verifies at once that φ is continuous and that $p \circ \varphi = p_0$, where $p: \mathcal{O}(S) \to \mathbb{C}^n$ is the natural projection. In particular, φ is a local isomorphism.

Since Ω is connected, so is $\varphi(\Omega)$. Let X be the connected component of $\mathcal{O}(S)$ containing $\varphi(\Omega)$, and denote again by p the restriction to X of the map $p: \mathcal{O}(S) \to \mathbb{C}^n$.

We claim that $p: X \to \mathbb{C}^n$ and $\varphi: \Omega \to X$ is an S-envelope of holomorphy of Ω.

First, we observe that, for all $s \in S$, we have a holomorphic function F_s on $\mathcal{O}(S)$ defined as follows. If $\underline{g}_z \in \mathcal{O}_z(S)$ is defined by $(V, \{g_s\})$, we set $F_s(\underline{g}_z) = g_s(z)$. One verifies at once that F_s is holomorphic on $\mathcal{O}(S)$. We denote the restriction of F_s to X again by F_s. Now, by the very definition of φ, it follows that $F_s \circ \varphi = s$ for all $s \in S$. This proves property (a) in Definition 1.

To prove (b), let $p': X' \to \mathbb{C}^n$, $\varphi': \Omega \to X'$ be given with $p' \circ \varphi' = p_0$, and suppose that for all $s \in S$, there exists $F'_s \in \mathcal{H}(X')$ so that $s = F'_s \circ \varphi'$. Let $S' = \{F'_s\}_{s \in S}$. Let $u: X' \to \mathcal{O}(S)$ be the map $\varphi(p', S')$ (defined at the beginning of the proof). Since $F'_s \circ \varphi' = s$ and $p' \circ \varphi' = p_0$, we have $\varphi = u \circ \varphi'$ (locally, $F'_s \circ p'^{-1} = F'_s \circ \varphi' \circ \varphi'^{-1} \circ p'^{-1} = s \circ p_0^{-1}$). Clearly, $p' = p \circ u$. This proves (b), and with it Theorem 1.

Definition 2. If $p_o : \Omega \to \mathbb{C}^n$ is a connected domain over $\mathbb{C}^n$, and $S = \mathcal{H}(\Omega)$, the S-envelope of holomorphy of Ω is called simply the envelope of holomorphy of Ω.

Proposition 2. Let $p_o : \Omega \to \mathbb{C}^n$ be a connected domain over $\mathbb{C}^n$ and $f \in \mathcal{H}(\Omega)$. Let F be its extension to the envelope of holomorphy $p : X \to \mathbb{C}^n$. Then $f(\Omega) = F(X)$. In particular, if f is bounded, $|f(x)| < M$ for all $x \in \Omega$, then F is bounded and $|F(x)| < M$ for all $x \in X$.

Proof. Since $f = F \circ \varphi$, we have $f(\Omega) \subset F(X)$. Suppose that there is $c \in F(X) - f(\Omega)$. Then $1/(f-c) \in \mathcal{H}(\Omega)$. Let G be its extension to X. Then $G \cdot (F-c)$ is the extension to X of $1 = (f-c)^{-1} \cdot (f-c)$, so that $G \cdot (F-c) \equiv 1$ on X. This implies that $F(x) \neq c$ for all $x \in X$, a contradiction.

Proposition 3. Let $p_o : \Omega \to \mathbb{C}^n$ and $p_o' : \Omega' \to \mathbb{C}^n$ be connected domains over $\mathbb{C}^n$, and $p : X \to \mathbb{C}^n$, $p' : X' \to \mathbb{C}^n$ their envelopes of holomorphy. Let $u : \Omega \to \Omega'$ be a holomorphic map which is a local isomorphism. Then, there exists a holomorphic map $\tilde{u} : X \to X'$ such that the diagram

$$\begin{array}{ccc} \Omega & \xrightarrow{\ u\ } & \Omega' \\ {\scriptstyle \varphi}\downarrow & & \downarrow{\scriptstyle \varphi'} \\ X & \xrightarrow{\ \tilde{u}\ } & X' \end{array}$$

commutes. Here $\varphi : \Omega \to X$, $\varphi' : \Omega' \to X'$ are the mappings of Definition 1.

Proof. Let $v = \varphi' \circ u : \Omega \to X'$. Then v is holomorphic and a local isomorphism. We have to show that there is $u : X \to X'$ so that $u \circ \varphi = v$. Consider the map $\psi = p' \circ v : \Omega \to \mathbb{C}^n$. Then ψ is again a local isomorphism. If $\psi = (\psi_1, \ldots, \psi_n)$, the ψ_j are holomorphic. Let η be the jacobian determinant $\eta = \det(\frac{\partial \psi_i}{\partial x_j})$, where $\frac{\partial f}{\partial x_j}$, for a holomoprhic f on Ω is as in chapter 2, Definition 5. Then, since ψ is a local isomorphism, $\eta(x) \neq 0$ for all $x \in \Omega$. Let Ψ_j be the extension of ψ_j to X, and let $\Psi = (\Psi_1, \ldots, \Psi_n)$. Let H be the extension of η to X. Then, clearly, $H = \det(\frac{\partial \Psi_i}{\partial x_j})$. Moreover, by Proposition 2, $H(x) \neq 0$ for all $x \in X$. Hence $\Psi: X \to \mathbb{C}^n$ is a local isomorphism (chapter 5, Lemma 3). Moreover, $\Psi \circ \varphi = \psi$.

Consider now the domains $\psi: \Omega \to \mathbb{C}^n$, $p': X' \to \mathbb{C}^n$ and $v: \Omega \to X'$. Let $S = \{f \circ u \mid f \in \mathcal{H}(\Omega')\} = \{F \circ v \mid F \in \mathcal{H}(X')\}$. We claim that $p': X' \to \mathbb{C}^n$ is the S-envelope of holomorphy of $\psi: \Omega \to \mathbb{C}^n$. (See the following lemma.) Now, any holomorphic function on Ω can be extended to X, so that $\Psi: X \to \mathbb{C}^n$ is an S-extension of $\psi: \Omega \to \mathbb{C}^n$ relative to the map $\varphi: \Omega \to X$. Since $p': X' \to \mathbb{C}^n$ is the S-envelope of holomorphy of $\psi: \Omega \to \mathbb{C}^n$, there is a holomorphic map $\tilde{u}: X \to X'$ such that $p' \circ \tilde{u} = \Psi$ and $\tilde{u} \circ \varphi = v$. This prove the proposition.

The fact that $p': X' \to \mathbb{C}^n$ is the S-envelope of holomorphy of $\psi: \Omega \to \mathbb{C}^n$ is a consequence of the following lemma.

<u>Lemma 1</u>. Let $p_0: \Omega \to \mathbb{C}^n$, $p'_0: \Omega' \to \mathbb{C}^n$ be connected domains over $\mathbb{C}^n$ and $p': X' \to \mathbb{C}^n$, $\varphi': \Omega' \to X'$ be the T-envelope of holomorphy of $p'_0: \Omega' \to \mathbb{C}^n$, $T \subset \mathcal{H}(\Omega')$. Let $u: \Omega \to \Omega'$ be a local holomorphic isomorphism, and let $S = \{f \circ u \mid f \in T\}$. Then $p': X' \to \mathbb{C}^n$, $\varphi' \circ u: \Omega \to X'$ is the S-envelope of holomorphy of $q_0: \Omega \to \mathbb{C}^n$ where $q_0 = p'_0 \circ u$.

Proof. We identify X' with a certain connected component of $\mathcal{O}(T)$ and the map φ' with the map $\varphi(p_0', T)$ constructed at the beginning of the proof of Theorem 1. The map $f \mapsto f \circ u$ is a bijection of T onto S, so that we may identify $\mathcal{O}(T)$ and $\mathcal{O}(S)$. We see at once that $\varphi(q_0, S)(\Omega) \subset \varphi(p_0', T)(\Omega')$, so that the connected component of $\mathcal{O}(T) = \mathcal{O}(S)$ containing $\varphi(q_0, S)(\Omega)$ is X'.

Corollary. If $p: X \to \mathbb{C}^n$, $\varphi: \Omega \to X$ is the envelope of holomorphy of $p_0: \Omega \to \mathbb{C}^n$, then $p: X \to \mathbb{C}^n$ is a domain of holomorphy, i.e., the natural map φ of X into its envelope of holomorphy is an isomorphism (see Definition 4 below).

Corollary to Proposition 3. Let $p_0: \Omega \to \mathbb{C}^n$ be a connected domain over $\mathbb{C}^n$ and $p: X \to \mathbb{C}^n$, $\varphi: \Omega \to X$ its envelope of holomorphy. Then, for any analytic automorphism σ of Ω (i.e., homeomorphism σ such that σ and σ^{-1} are holomorphic), there exists an analytic automorphism $\tilde{\sigma}$ of X such that $\varphi \circ \sigma = \tilde{\sigma} \circ \varphi$.

Proof. By Proposition 2, there is a holomorphic map $\tilde{\sigma}: X \to X$ such that $\varphi \circ \sigma = \tilde{\sigma} \circ \varphi$. Also there is a holomorphic map $\tau: X \to X$ such that $\varphi \circ \sigma^{-1} = \tau \circ \varphi$. Also $\tau \circ \tilde{\sigma} \circ \varphi = \tau \circ \varphi \circ \sigma = \varphi \circ \sigma^{-1} \circ \sigma = \varphi$, so that $\tau \circ \tilde{\sigma} =$ identity on $\varphi(\Omega)$, hence on X. Similarly, $\tilde{\sigma} \circ \tau$ is the identity and the result follows.

Proposition 4. Let $p_0: \Omega \to \mathbb{C}^n$ be a connected domain over $\mathbb{C}^n$ and let $q_0: \Omega \to \mathbb{C}^n$ be a holomorphic map (relative to p_0) that is a local isomorphism. Let $p: X \to \mathbb{C}^n$, $\varphi: \Omega \to X$, $q: Y \to \mathbb{C}^n$, $\psi: \Omega \to Y$ be the envelopes of holomorphy of $p_0: \Omega \to \mathbb{C}^n$ and $q_0: \Omega \to \mathbb{C}^n$. Then there is an analytic isomorphism $F: X \to Y$ such that $\psi = F \circ \varphi$.

morphism.

Definition 4. Let $p_o:\Omega \to \mathbb{C}^n$ be a connected domain over $\mathbb{C}^n$ and $S \subset \mathcal{H}(\Omega)$. Ω is called an S-domain of holomorphy (or the domain of existence of S) if the natural map of Ω into its S-envelope of holomorphy is an analytic isomorphism. If $S = \mathcal{H}(\Omega)$, Ω is called simply a domain of holomorphy.

Note that, by the corollary above, the property of being a domain of holomorphy is independent of the local analytic isomorphism $p_o:\Omega \to \mathbb{C}^n$ (so long as the different maps used are local analytic isomorphisms relative to each other).

Lemma 2. Let $p_o:\Omega \to C^n$ be a connected domain, $S \subset \mathcal{H}(\Omega)$. Suppose that for any $f \in S$ and $\alpha \in \mathbb{N}^n$, we have $D^\alpha f \in S$. Let $p:X \to \mathbb{C}^n$, $\varphi:\Omega \to X$ be the S-envelope of holomorphy of p_o. Then φ is injective if and only if S separates points of $p_o^{-1} p_o(a)$ for any $a \in \Omega$.

Proof. We identify X with a connected component of $\mathcal{O}(S)$, φ with $\varphi(p_o, S)$ as in the proof of Theorem 1. It is clear that φ is injective if and only if the following holds:

Let $a, a' \in p_0^{-1}(x_0)$, $x_0 \in p_0(\Omega)$, $a \neq a'$ and let U, U' be neighborhoods of a, a' respectively such that $p_0|U = q_0$, $p_0|U' = q_0'$ are isomorphisms of U, U' onto a polydisc P about x_0. Then, there is $f \in S$ such that $f \circ q_0^{-1} \not\equiv f \circ q_0'^{-1}$ on P. This is the case if and only if there is $\alpha \in \mathbb{N}^n$ such that

$$(D^\alpha f) \circ q_0^{-1}(x_0) \neq (D^\alpha f) \circ q_0'^{-1}(x_0)$$

and the result follows.

Corollary 1. If $S = \mathcal{H}(\Omega)$, then if φ is injective, $\mathcal{H}(\Omega)$ separates points of Ω.

Proof. Let $p_0 = (p_1, \dots, p_n)$, where $p_j \in \mathcal{H}(\Omega)$. Let $a, a' \in \Omega$, $a \neq a'$. If $p_0(a) = p_0(a')$, by Lemma 2, there is $f \in S$ with $f(a) \neq f(a')$. If $p_0(a) \neq p_0(a')$, there is j, $1 \leq j \leq n$, with $p_j(a) \neq p_j(a')$.

Corollary 2. If $p: X \to \mathbb{C}^n$ is the envelope of holomorphy of $p_0: \Omega \to \mathbb{C}^n$, then $\mathcal{H}(X)$ separates points of X.

Examples. We shall now give an example to show that the consideration of domains over $\mathbb{C}^n$ is necessary; there are domains in $\mathbb{C}^n$ whose envelopes of holomorphy are no longer in $\mathbb{C}^n$. The following example is due to H. Cartan.

Let $(z, w) \in \mathbb{C}^2$, $z = x + iy$. Let

$$\Omega_1 = \{(z, w) \in \mathbb{C}^2 \mid -4 < x < 0,\ |w| < e^x\},$$

$$\Omega_2 = \{(z, w) \in \mathbb{C}^2 \mid 0 \leq x < 4,\ e^{-1/x} < |w| < 1\},$$

and let $\Omega = \Omega_1 \cup \Omega_2$. Then Ω is a connected open subset of $\mathbb{C}^n$. Let $p_0: \Omega \to \mathbb{C}^n$ be the map $p_0(z, w) = (\exp(iz), w)$. Then p_0 is injective, since, if $p_0(z, w) = p_0(z', w')$, then $w = w'$ and $\operatorname{Im} z = \operatorname{Im} z'$,

$\operatorname{Re} z' - \operatorname{Re} z = 2k\pi$, $k \in \mathbb{Z}$. Suppose that $\operatorname{Re} z < \operatorname{Re} z'$, (i.e., $k \geq 1$). Then, since $\operatorname{Re} z' < 4$, $\operatorname{Re} z < 4 - 2\pi$. We see that $k = 1$ and that

$$e^{-1/\operatorname{Re} z'} < |w| < e^{\operatorname{Re} z} ,$$

that is,

$$e^{-\frac{1}{x+2\pi}} < |w| < e^{x} \quad , \ x = \operatorname{Re} z .$$

Since $x > -4$, $x + 2\pi > 2$, so that we have, in particular,

$$e^{-1/2} < e^{4-2\pi} ,$$

which is absurd.

Now if f is holomorphic in Ω, then f can be expanded in a series

$$f(z, w) = \sum_{\nu=-\infty}^{\infty} a_\nu(z) w^\nu$$

converging uniformly on compact subsets of Ω. If we fix z with $\operatorname{Re} z < 0$, then $w \mapsto f(z, w)$ is holomorphic in a neighborhood of $w = 0$. Hence $a_\nu(z) = 0$ for $\nu < 0$ and $\operatorname{Re} z < 0$. By analytic continuation, $a_\nu(z) \equiv 0$, $z \in \Omega$, $\nu < 0$. Hence

$$f(z, w) = \sum_{\nu=0}^{\infty} a_\nu(z) w^\nu .$$

Since this series converges uniformly on compact subsets of Ω, it converges uniformly on compact subsets of

$$X = \Omega_1 \cup \Omega_2' \quad , \qquad \Omega_2' = \{0 \leq x < 4 , \ |w| < 1\}$$

Hence $p: X \to \mathbb{C}^n$, $p(z, w) = (\exp(iz), w)$, is an $\mathcal{H}(\Omega)$-extension of $p_0: \Omega \to \mathbb{C}^n$. But p is not injective (e.g., $p(-\pi, e^{-2\pi}) = p(\pi, e^{-2\pi})$). It follows at once that the envelope of holomorphy has the same property. [Actually, $p: X \to \mathbb{C}^n$ is the envelope of holomorphy of $p_o: \Omega \to \mathbb{C}^n$.]

Another example of this kind is the following.

Let $H = \{z \in \mathbb{C} \mid \operatorname{Re} z > 0\}$ and $\Omega_1 = H \times \mathbb{C} \subset \mathbb{C}^2$. Let $\Omega_2 = \Omega_1 - \{(z,w) \in \mathbb{C}^2 \mid |z+3| \leq 2,\ |w| \leq 2\}$, $\Omega = \Omega_2 \cup \{(z,w) \in \mathbb{C}^2 \mid |z-3| < 1,\ |w| < 1\}$. Let $p_o: \Omega \to \mathbb{C}^n$ be the map $(z,w) \mapsto (z^2, w)$. It follows easily from the results of chapter 2 that any holomorphic function on Ω_2 can be continued to Ω_1, and the map $(z,w) \mapsto (z^2, w)$ is not injective on $\Omega_1 \cup \{(z,w) \in \mathbb{C}^2 \mid |z-3| < 1,\ |w| < 1\}$.

We shall next show that there are domains $p_o: \Omega \to \mathbb{C}^n$ over $\mathbb{C}^n$, for which p_o is not injective, such that, nevertheless, the projection $p: X \to \mathbb{C}^n$ of the envelope of holomorphy is injective. We shall need the following theorem.

Let $p_o: \Omega \to \mathbb{C}^n$ be a connected domain over $\mathbb{C}^n$. We say that the domain is a Reinhardt domain if the following holds.

There is given $a_o \in \Omega$ with $p_o(a_o) = 0$ (called the origin of Ω)

Let $T^n = \{(\zeta_1, \ldots, \zeta_n) \in \mathbb{C}^n \mid |\zeta_1| = 1, \ldots, |\zeta_n| = 1\}$. For any $\zeta \in T^n$ is given an analytic automorphism σ_ζ of Ω such that $p_o \circ \sigma_\zeta = \zeta \cdot p_o$ and such that $\sigma_\zeta(a_o) = a_o$ for all $\zeta \in T^n$. Here $(\zeta \cdot p_o(x) = (\zeta_1 p_1(x), \ldots, \zeta_n p_n(x))$ if $p_o = (p_1, \ldots, p_n)$. One verifies at once that $\sigma_\zeta \circ \sigma_{\zeta'} = \sigma_{\zeta \cdot \zeta'}$ (where $\zeta \cdot \zeta' = (\zeta_1 \zeta_1', \ldots, \zeta_n \zeta_n')$ if $\zeta = (\zeta_1, \ldots, \zeta_n)$, $\zeta' = (\zeta_1', \ldots, \zeta_n')$). Moreover, it can be shown that the map $T^n \times \Omega \to \Omega$ given by $(x, \zeta) \mapsto \sigma_\zeta(x)$ is continuous.

Under these hypotheses we have the following theorem.

Proposition 5. For any $f \in \mathcal{H}(\Omega)$, there is a power series $\sum_{\alpha \in \mathbb{N}^n} c_\alpha T^\alpha$ such that

$$f(x) = \sum_{\alpha \in \mathbb{N}^n} c_\alpha (p_o(x))^\alpha$$

and the series converges uniformly on compact subsets of Ω.

Proof. The proof is essentially the same as that of chapter 2, Theorem 2. Let $a \in \Omega$ and U a neighborhood of a such that $p_o|U$ is an isomorphism onto a polydisc P about $p_o(a)$. For $x \in \Omega$, let $p_o(x) = (x_1, \ldots, x_n)$ and let $x_{i_1}, \ldots, x_{i_k}$, $1 \le i_1 < \ldots < i_k \le n$ be those coordinates of $p_o(x)$ which are 0. If $\zeta \in T^n$, we set $\zeta' = (\zeta'_1, \ldots, \zeta'_n)$ where $\zeta'_j = \zeta_j$ if $j \neq i_p$, $p = 1, \ldots, k$, and $\zeta'_{i_p} = 1$, $p = 1, \ldots, k$. Then $(\zeta_1 x_1, \ldots, \zeta_n x_n) = (\zeta'_1 x_1, \ldots, \zeta'_n x_n)$, so that the mappings $\zeta \mapsto \sigma_\zeta(x)$, $\zeta \mapsto \sigma_{\zeta'}(x)$ are two liftings of the map $\zeta \mapsto \zeta \cdot p_o(x)$. Since they coincide for $\zeta = (1, \ldots, 1)$, it follows that $\sigma_\zeta(x) = \sigma_{\zeta'}(x)$ for all $\zeta \in T^n$.

We claim now that if $T^n(x) = \bigcup_{\zeta \in T^n} \{\sigma_\zeta(x)\}$, then $p_o|T^n(x)$ is injective. In fact, if $p_o(\sigma_\zeta(x)) = p_o(\sigma_\eta(x))$, we have $\zeta \cdot p_o(x) = \eta p_o(x)$, hence $\zeta' \cdot p_o(x) = \eta' p_o(x)$ and, since the coordinates of ζ', η' corresponding to $i_1, \ldots, i_k$ are equal, it follows that $\zeta' = \eta'$. Hence

$$\sigma_\zeta(x) = \sigma_{\zeta'}(x) = \sigma_{\eta'}(x) = \sigma_\eta(x).$$

It follows easily that $T^n(x)$ has a neighborhood N such that $p_o|N$ is injective. Hence, if U above is small enough, then $p_o | \bigcup_{\zeta \in T^n} \sigma_\zeta(U)$ is an isomorphism onto $\bigcup_{\zeta \in T^n} \zeta \cdot P$. Now, this latter set is of the form $\{x \in \mathbb{C}^n | \ r_j < |z_j| < R_j\}$. Hence, it follows that for any

$f \in \mathcal{H}(\Omega)$, there is a series $\sum_{\alpha \in \mathbb{Z}^n} c_\alpha T^\alpha$ such that $\sum c_\alpha (p_o(x))^\alpha$ converges to f uniformly on a neighborhood of a. As in chapter 2, Theorem 2, there is a series $\sum_{\alpha \in \mathbb{Z}^n} c_\alpha (p_o(x))^\alpha$ which converges to f(x) on any compact subset of Ω. If we look at a neighborhood of the origin $a_o \in \Omega$, we see that this series must be of the form

$$\sum_{\alpha \in \mathbb{N}^n} c_\alpha (p_o(x))^\alpha .$$

Proposition 6. Let $p_o:\Omega \to \mathbb{C}^n$ be a Reinhardt domain and let $p:X \to \mathbb{C}^n$, $\varphi:\Omega \to X$ be its envelope of holomorphy. Then X is a Reinhardt domain and $p:X \to \mathbb{C}^n$ is injective.

Proof. Let $x_o = \varphi(a_o)$. Let $\zeta \to \sigma_\zeta$ be the map of T^n into the automorphisms of Ω defining the structure of Reinhardt domain on Ω. By Corollary 1 to Proposition 3, there is an automorphism τ_ζ of X such that $\varphi \circ \sigma_\zeta = \tau_\zeta \circ \varphi$. In particular, $\tau_\zeta(x_o) = x_o$. Clearly the τ_ζ make of X a Reinhardt domain.

By Proposition 5, any $f \in \mathcal{H}(X)$ takes the same value at any two points of $p^{-1}p(x)$, $x \in X$. By Corollary 2 to Lemma 2, $\mathcal{H}(X)$ separates points of X. Hence $p^{-1}p(x)$ reduces to the single point x for $x \in X$, so that p is injective.

Thus, to construct a domain not in $\mathbb{C}^n$ whose envelope of holomorphy is in $\mathbb{C}^n$, it is sufficient to construct a Reinhardt domain $p_o:\Omega \to \mathbb{C}^n$ for which p_o is not injective.

Let Q_o be the following set in $\mathbb{R}^2$:

$$Q_o = \{(x,y) \in \mathbb{R}^2 \mid 0 \le x < 2,\ 0 \le y < 2\} - \{(x,y) \in \mathbb{R}^2 \mid x = 1,\ 0 \le y \le 1\}.$$

Let

$$Q_1 = \{(x, y) \in \mathbb{R}^2 \mid 0 \le x < 2,\ 0 \le y < 1\},$$

and let $D_j = \{(z,w) \in \mathbb{C}^2 \mid (|x|, |w|) \in Q_j\}$, $j = 0, 1$. Consider the <u>disjoint</u> union $X = D_0 \cup D_1$ and introduce the following equivalence relation in X: $(z_0, w_0) \sim (z_1, w_1)$ if and only if $z_0 = z_1$, $w_0 = w_1$ and $1 < |z_1| < 2$, $0 \le |w_1| < 1$; here $(z_0, w_0) \in D_0$, $(z_1, w_1) \in D_1$. Let Ω be the quotient of X by this relation and $p_0 : \Omega \to \mathbb{C}^2$ the map induced by the inclusion of D_0, D_1 into $\mathbb{C}^2$. Then $p_0 : \Omega \to \mathbb{C}^2$ is clearly a Reinhardt domain for which p_0 is not injective.

For references related to the results of this chapter, see [1], [10], [19], [29].

7
DOMAINS OF HOLOMORPHY: CONVEXITY THEORY

Throughout this chapter, $p_o : \Omega \to \mathbb{C}^n$ will be a connected domain over $\mathbb{C}^n$.

Definition 1. (a) If $f \in \mathcal{H}(\Omega)$ and A is a subset of Ω, we write

$$\|f\|_A = \sup_{x \in A} |f(x)|. \qquad (\leq \infty)$$

(b) If A is a subset of Ω, and $S \subset \mathcal{H}(\Omega)$, we set

$$\hat{A}_S = \{x \in \Omega \mid |f(x)| \leq \|f\|_A \text{ for all } f \in S\}.$$

If $S = \mathcal{H}(\Omega)$, we write $\hat{A} = \hat{A}_S$.

Remark. If S is closed under multiplication, then

$$\hat{A}_S = \{x \in \Omega \mid \text{there exists } M_x > 0 \text{ such that } |f(x)| \leq M_x \|f\|_A \text{ for all } f \in S\}.$$

Proof. If we denote by B the set on the right, we have $\hat{A}_S \subset B$. Now, if $x \in B$, and $f \in S$, then $f^p \in S$, $p = 1, 2, \ldots$, so that

$$|f(x)|^p \leq M_x \|f\|_A^p .$$

Since $M_x^{1/p} \to 1$ as $p \to \infty$, it follows that $|f(x)| \leq \|f\|_A$, and $x \in \hat{A}_S$.

Lemma 1. Let K be compact and $M > 0$, $\varepsilon > 0$. Then, for $a \notin \hat{K}$, there exists $f \in \mathcal{H}(\Omega)$ such that

$$f(a) = M , \quad \|f\|_K < \varepsilon .$$

Proof. Let $g \in \mathcal{H}(\Omega)$ be such that $|g(a)| > \|g\|_K$. Then, we may take $f = M(g/g(a))^p$ with a large enough integer p.

Lemma 2. Let $A \subset \Omega$ be a subset such that $\|f\|_A < \infty$ for any $f \in \mathcal{H}(\Omega)$. Then, there is a compact set $K \subset \Omega$ such that $A \subset \hat{K}$.

Proof. Suppose the result false. Let $\{K_p\}_{p=0,1,\ldots}$ be a sequence of compact sets in Ω with $K_p \subset \overset{\circ}{K}_{p+1}$, $\bigcup K_p = \Omega$. Then, since $A \not\subset \hat{K}_p$, there is $x_p \in A - \hat{K}_p$. By replacing $\{K_p\}$ by a subsequence, if necessary, we may suppose that $x_p \in K_{p+1}$. Let $f_o \in \mathcal{H}(\Omega)$ be such that

$$|f_o(x_o)| > 1, \ \|f_o\|_{K_o} < 1,$$

and, by induction, let $f_p \in \mathcal{H}(\Omega)$ be such that

$$(1) \qquad |f_p(x_p)| > p+1 + \sum_{q=0}^{p-1} |f_q(x_p)| \ , \ \|f_p\|_{K_p} < 2^{-p} .$$

Then the series $\sum_{p=0}^{\infty} f_p = f$ converges uniformly on every compact subset of Ω, so that $f \in \mathcal{H}(\Omega)$. Moreover

$$|f(x_p)| \geq |f_p(x_p)| - \sum_{q=p+1}^{\infty} |f_q(x_p)| - \sum_{q=0}^{p-1} |f_q(x_p)|$$

$$\geq p+1 - \sum_{q=p+1}^{\infty} |f_q(x_p)|$$

by (1). Now, for $q > p$, $x_p \in K_q$. Hence $|f_q(x_p)| \leq \|f_q\|_{K_q} < 2^{-q}$. Hence $\sum_{q=p+1}^{\infty} |f_q(x_p)| < 1$, so that $|f(x_p)| > p$. Since $x_p \in A$, $p = 0, 1, 2, \ldots$ f is not bounded on A, hence $\|f\|_A = \infty$, contradicting our hypothesis.

Lemma 3. The following two statements are equivalent.

(a) For any $K \subset \Omega$, K compact, $\hat{K}$ is also compact.

(b) For any (infinite) sequence $\{x_\nu\}$ which has no limit point in Ω, there exists $f \in \mathcal{H}(\Omega)$ such that $\{f(x_\nu)\}$ is unbounded.

Proof. (a) $\Rightarrow$ (b). Let $\{x_\nu\}$ be a sequence without limit point in Ω. Then $\{x_\nu\} \not\subset \hat{K}$ for any compact K. By Lemma 2, there is $f \in \mathcal{H}(\Omega)$ such that $\{f(x_\nu)\}$ is not bounded.

(b) $\Rightarrow$ (a). If $\hat{K}$ is not compact, there exists a sequence $\{x_\nu\}$, $x_\nu \in \hat{K}$, which has no limit point in Ω. Let $f \in \mathcal{H}(\Omega)$ be such that $\{f(x_\nu)\}$ is unbounded. Then $\|f\|_{\hat{K}} = \infty$. But, it follows from the definition of $\hat{K}$ that $\|f\|_{\hat{K}} = \|f\|_K < \infty$.

If the conditions of Lemma 3 are satisfied, we say that Ω is holomorphically convex.

Definition 2. Let $a \in \Omega$. A polydisc of radius r about a is a connected open set U, $a \in U$ such that $p_0|U$ is an analytic isomorphism onto the set $\{z \in \mathbb{C}^n \mid |z_j - b_j| < r\}$; here $p_0(a) = (b_1, \ldots, b_n)$. We denote the set U by $P(a, r)$. The maximal polydisc $P(a, r_0)$ is the union of all polydiscs about a.

Lemma 4. $P(a, r_0)$ is a polydisc about a of radius

$$r_0 = \sup r$$

where the supremum is over all polydiscs $P(a, r)$ about a.

Proof. It suffices to show that the map

$$p_0 : P(a, r_0) \to P = \{z \in \mathbb{C}^n \mid |z_j - b_j| < r_0\}$$

is bijective. [Clearly $p_0(P(a, r_0)) \subset P$.] p_0 is injective: if $x, x' \in P(a, r_0)$, there is a polydisc $P(a, r)$ containing both x and x', so that $p_o(x) \neq p_o(x')$. p_o is surjective: if $z \in P$, then $\max_j |z_j - b_j| < r_o$, hence there is a polydisc $P(a, r)$ of radius r, $|z_j - b_j| < r \leq r_o$, so that there is a point $x \in P(a, r)$ with $p_o(x) = z$.

Definition 3. The radius of the maximal polydisc about a is called the distance of a from the boundary of Ω and is denoted by $d(a)$ (or $d_{p_o}(a)$ when the dependence on $p_o: \Omega \to \mathbb{C}^n$ is relevant).

Lemma 5. If there is a point $a \in \Omega$ with $d(a) = \infty$, then p_o is an isomorphism of Ω onto $\mathbb{C}^n$.

Proof. To say that $d(a) = \infty$ means simply that there is an open set U containing a such that $p_o|U$ is an isomorphism onto $\mathbb{C}^n$. It follows at once that $\{x \in \Omega \mid d(x) = \infty\}$ is open. Moreover, it is closed (if $x_\nu \in \Omega$, $x_\nu \to x_o$ and P is a polydisc about x_o, and if U is a neighborhood of x_ν, $x_\nu \in P$, such that $p_o|U$ is an isomorphism onto $\mathbb{C}^n$, then $x_o \in U$ and $d(x_o) = \infty$). Hence $d(x) = \infty$ for any $x \in \Omega$. It follows at once that p_o is a covering. Since $\mathbb{C}^n$ is simply connected, $p_o: \Omega \to \mathbb{C}^n$ is an isomorphism.

Remark. One can prove that if there is $\rho > 0$ such that $d(x) \geq \rho$ for all $x \in \Omega$, then p_o is an isomorphism onto $\mathbb{C}^n$.

Lemma 6. If $d < \infty$, then the function $a \to d(a)$ is continuous on Ω.

Proof. Let $a \in \Omega$ and P the maximal polydisc, of radiux $\rho > 0$, about a. Let U be the polydisc of radius $\rho/4$ about a. Clearly, if $x \in U$, $p_o(P)$ contains the polydisc about $p_o(x)$ of radius $\rho - |p_o(x) - p_o(a)|$, hence P contains a polydisc about x of radius $\rho - |p_o(x) - p_o(a)|$ about x. Hence

$$d(x) \geq d(a) - |p_o(x) - p_o(a)|.$$

Similarly, $d(a) \geq d(x) - |p_o(a) - p_o(x)|$, so that

$$|d(x) - d(a)| \leq |p_o(x) - p_o(a)| \quad \text{for } x \in U.$$

Definition 4. If A is a subset of Ω, we set

$$d(A) = \inf_{a \in A} d(a) .$$

Since d is continuous, if K is a compact set, then $d(K) > 0$.

Lemma 7. Let $a \in \Omega$, and $P = P(a, r_o)$, $r_o = d(a)$, be the maximal polydisc about a. Then $d(P) = 0$.

Proof. We begin with the following remark. Let $P = P(x, r)$ and $P' = P(x', r')$ be two polydiscs in Ω. Let
$Q = p_o(P) = \{z \in \mathbb{C}^n | \ |z-p_o(x)| < r\}$, $Q' = p_o(P') = \{z \in \mathbb{C}^n | \ |x-p_o(x')| < r'\}$.
Then, if $Q \cap Q' \neq \emptyset$, either $P \cap P' = \emptyset$ or p_o maps $P \cap P'$ isomorphically onto $Q \cap Q'$. In fact, let $q = (p_o|P)^{-1}$ and $q' = (p_o|P')^{-1}$
Then $q|Q \cap Q'$, $q'|Q \cap Q'$ are two liftings of the identity map of $Q \cap Q'$ (which is connected) and coincide on $p_o(P \cap P')$, hence everywhere by chapter 2, Lemma 4.

Let now $a \in \Omega$, $P = P(a, r_o)$, $r_o = d(a)$. Suppose that $d(P) = \rho > 0$. Let $x \in P$ and $P(x)$ the polydisc of radius ρ about x, and let $U = \bigcup_{x \in P} P(x)$. We claim that $p_o|U$ is an isomorphism onto $p_o(U)$. It suffices to show that $p_o|U$ is injective. Let $y \in P(x)$, $y' \in P(x')$ and suppose that $p_o(y) = p_o(y')$. Let $Q(x) = p_o(P(x))$, $Q(x') = p_o(P(x'))$. Then $p_o(y) = p_o(y') \in Q(x) \cap Q(x')$. Moreover, $P(x) \cap P$, $P(x') \cap P$ are mapped isomorphically onto $Q(x) \cap Q$, $Q(x') \cap Q$ respectively, where $Q = p_o(P)$.

Clearly, if $Q(x) \cap Q(x') \neq \emptyset$, then $Q(x) \cap Q(x') \cap Q \neq \emptyset$. Hence $P(x) \cap P(x') \cap P \neq \emptyset$. Hence, by our remark above, p_0 maps $P(x) \cap P(x')$ isomorphically onto $Q(x) \cap Q(x')$. This implies that $y, y' \in P(x) \cap P(x')$. Since $p_0(y) = p_0(y')$, and $p_0 | P(x) \cap P(x')$ is injective, this implies that $y = y'$, so that $p_0 | U$ is injective.

Now, $p_0(U) = \bigcup_{z \in Q} Q_z$, where $Q_z = \{w \in \mathbb{C}^n | \ |w-z| < \rho\}$, so that $p_0(U)$ contains a polydisc about $p_0(a)$ of radius $R > r_0$. This implies that $d(a) \geq R > r_0$, a contradiction.

Thus $d(P) = 0$.

Proposition 1. Let K be a compact subset of Ω and $x_0 \in \hat{K}$. Let $a = p_0(x_0)$, and let V be a polydisc about x_0, and $P = p_0(V)$. Then, for any $f \in \mathcal{H}(\Omega)$, if $g = f \circ (p_0 | V)^{-1} \in \mathcal{H}(P)$, the series

$$\sum_{\alpha \in \mathbb{N}^n} \frac{D^\alpha g(a)}{\alpha!} (z-a)^\alpha$$

converges in the polydisc $\{z \in \mathbb{C}^n | \ |z-a| < d(K)\}$.

Proof. Let $0 < r < d(K)$. For any $x \in K$, let $Q(x)$ be the polydisc of radius r about x and let $K' = \bigcup_{x \in K} \overline{Q(x)}$. Then K' is compact. Let $M = \|f\|_{K'}$. By Cauchy's inequality applied to $f | Q(x)$, we have $|D^\alpha f(x)| \leq M \cdot \alpha! \, r^{-|\alpha|}$, $x \in K$, so that $\|D^\alpha f\|_K \leq M \cdot \alpha! \, r^{-|\alpha|}$. Hence, by definition of $\hat{K}$, we have

$$\|D^\alpha f\|_{\hat{K}} \leq M \cdot \alpha! \, r^{-|\alpha|} .$$

Since $x_0 \in \hat{K}$, this implies that $|D^\alpha g(a)| \leq M \cdot \alpha! \, r^{-|\alpha|}$. It follows that the series $\sum D^\alpha g(a) \cdot (\alpha!)^{-1} (z-a)^\alpha$ converges for $|z-a| < r$. Since $r < d(K)$ is arbitrary, the result follows.

Theorem 1 (H. Cartan - P. Thullen). Suppose that $p_0 : \Omega \to \mathbb{C}^n$ is a domain of holomorphy. Then, for any compact set $K \subset \Omega$, we have

$$d(K) = d(\hat{K}).$$

Proof. Clearly $d(K) \geq d(\hat{K})$. Suppose that strict inequality holds. Then, there is $x_0 \in \hat{K}$ such that $d(x_0) = r_0 < \rho = d(K)$. Let $a = p_0(x_0)$ and let $P_0 = \{z \in \mathbb{C}^n \mid |z-a| < d(K) = \rho\}$. Let Ω_0 be the connected component of $p_0^{-1}(P_0)$ containing x_0. We assert that $p_0|\Omega_0$ is injective. In fact, it follows from Proposition 1 that for any $f \in \mathcal{H}(\Omega)$, there is $g_f \in \mathcal{H}(P_0)$ such that $f|\Omega_0 = g \circ p_0$. On the other hand, since $p_0 : \Omega \to C^n$ is a domain of holomorphy, it follows from chapter 6, Corollary 1 to Lemma 2, that $\mathcal{H}(\Omega)$ separates points of Ω. Since $g_f \circ p_0$ takes the same value at any two points $x, y \in \Omega_0$ with $p_0(x) = p_0(y)$, it follows that $p_0|\Omega_0$ is injective.

Let $r_0 < r < \rho$ and $\Omega_1 = \{x \in \Omega_0 \mid |p_0(x) - a| < r\}$, and let Y be the disjoint union of Ω and P, where $P = \{z \in \mathbb{C}^n \mid |z-a| < r\}$. We define an equivalence relation on Y by the requirement that $z \in P$ is equivalent to at most one point $x \in \Omega$, and that, if and only if $x \in \Omega_1$ and $p_0(x) = z$. Let X be the quotient of Y by this equivalence relation. Then X is Hausdorff, and the map of Y into $\mathbb{C}^n$ which is p_0 on Ω and the inclusion of P in $\mathbb{C}^n$ induces a local homeomorphism $p : X \to \mathbb{C}^n$. Moreover, the inclusion of Ω in Y induces a map $\varphi : \Omega \to X$ such that $p \circ \varphi = p_0$. We claim that for any $f \in \mathcal{H}(\Omega)$, there is $F_f \in \mathcal{H}(X)$ such that $F_f \circ \varphi = f$. In fact, since there is $g_f \in \mathcal{H}(P_0)$ such that $f|\Omega = g_f \circ p_0$ (Proposition 1), we define a function G_f on Y by $G_f|\Omega = f$, $G_f|P = g_g|P$. This induces $F_f \in \mathcal{H}(X)$ and we clearly have $F_f \circ \varphi = f$. Hence $p : X \to \mathbb{C}^n$, $\varphi : \Omega \to X$ is an $\mathcal{H}(\Omega)$-extension of

$p_o:\Omega \to \mathbb{C}^n$. Since $p_o:\Omega \to \mathbb{C}^n$ is, by assumption, a domain of holomorphy, it follows that φ is an analytic isomorphism. In particular, since X contains a polydisc of radius r about $\varphi(x_o)$, Ω contains a polydisc of radius r about x_o, contradicting our assumption that $r > r_o = d(x_o)$. This proves the theorem.

The same reasoning can be used to prove the following:

Theorem 1' (Cartan-Thullen). Let $S \subset \mathcal{H}(\Omega)$ be a subalgebra of $\mathcal{H}(\Omega)$ containing the functions $p_1, \ldots, p_n$, $(p_o = (p_1, \ldots, p_n))$ and closed under differentiation (i.e., $f \in S \Rightarrow D^\alpha f \in S$ for all $\alpha \in \mathbb{N}^n$). Then, if the natural map of Ω into its S-envelope of holomorphy is an isomorphism, we have $d(K) = d(\hat{K}_S)$ for any compact $K \subset \Omega$.

Corollary. If Ω is an open set in $\mathbb{C}^n$ which is a domain of holomorphy and p_o is the inclusion of Ω in $\mathbb{C}^n$, then for any compact set $K \subset \Omega$, $\hat{K}$ is also compact.

Proof. $\hat{K}$ is clearly closed in Ω. Moreover, since $d(K) = d(\hat{K})$, it follows that $\hat{K}$ is closed in $\mathbb{C}^n$ (since the closure of $\hat{K}$ in $\mathbb{C}^n$ cannot meet $\partial\Omega$). Moreover, $\hat{K}$ is contained in the polydisc $\{z \in \mathbb{C}^n \mid |z| \leq \rho\}$ where $\rho = \max_j \|z_j\|_K$, and so is bounded. Hence $\hat{K}$ is compact.

Theorem 2 (Cartan-Thullen). Let $p_o:\Omega \to \mathbb{C}^n$ have the property that for any compact set $K \subset \Omega$, we have $d(\hat{K}) > 0$. Suppose further that $\mathcal{H}(\Omega)$ separates points of Ω. Then, there is $g \in \mathcal{H}(\Omega)$ such that if we denote by $p:X \to \mathbb{C}^n$, $\varphi:\Omega \to X$ the S-envelope of holomorphy of $p_o:\Omega \to \mathbb{C}^n$ where $S = \{g\}$ is the set consisting of the single element g, then φ is an isomorphism.

In other words, Ω is the domain of existence of g.

Before starting on the proof, we give a definition.

Definition 5. Let $f \in \mathcal{H}(\Omega)$, $f \not\equiv 0$. Then, if $a \in \Omega$, the order of the zero of f at a is defined to be the largest integer $k \geq 0$ such that $D^{\alpha}f(a) = 0$ for all $\alpha \in \mathbb{N}^n$ with $|\alpha| < k$. We denote this by $\omega(f, a)$.

Note that the function $a \mapsto \omega(f, a)$ is upper semi-continuous on Ω. In particular, it is bounded above on any compact subset of Ω.

Proof of Theorem 2. Part 1: We shall prove the following. Let $\{x_\nu\}$ be a dense sequence in Ω and let P_ν be the maximal polydisc about x_ν. Then, there is $g \in \mathcal{H}(\Omega)$ such that

(a) g has zeros of arbitrarily large order in each P_ν

(b) There is a dense set $E \subset \Omega$ such that $p_o^{-1} p_o(E) = E$ and such that g separates the points of E.

To prove this, we proceed as follows. We consider the sequence

$$P_1, \; P_1, P_2, \; P_1, P_2, P_3, \; P_1, P_2, P_3, P_4, \ldots$$

(the essential property being that each P_k occurs in this sequence infinitely often). We denote by Q_p the p-th polydisc of this sequence. Let $\{K_p\}$ be a sequence of compact subsets of Ω such that $K_p \subset \overset{o}{K}_{p+1}$ and $\bigcup K_p = \Omega$. Then $d(\hat{K}_p) > 0$ by hypothesis. Hence, by Lemma 7, $Q_p \not\subset \hat{K}_p$. Let $y_p \in Q_p - \hat{K}_p$. By Lemma 1, there is $f_p \in \mathcal{H}(\Omega)$ such that

$$F_p(y_p) = 1, \quad \|F_p\|_{K_p} < 2^{-p}.$$

Let

(2) $$h(x) = \prod_{p=1}^{\infty} (1 - F_p(x))^p.$$

This product converges uniformly on each K_p since, for $q > p$, and

$x \in K_p$,

$$|(1 - F_q(x))^q - 1| \leq q\|F_q\|_{K_p} < q2^{-q}.$$

Since $\bigcup \overset{\circ}{K}_p = \Omega$, any compact subset of Ω is contained in a K_p, so that $f \in \mathcal{H}(\Omega)$. Furthermore, $h \not\equiv 0$ (since e.g., $|F_p(x)| < \frac{1}{2}$ for all p if $x \in K_1$, so that no term in this absolutely convergent product is 0 at $x \in K_1$) and h has a zero of order at least p at y_p. Since each $P_\nu = Q_p$ for infinitely many values of p, it follows that h has zeros of arbitrarily large order in each P_ν.

Let $A = \{x \in \Omega | \; h(x) = 0\}$. Then A is a closed nowhere dense set in Ω. Let $B = p_o(A) \subset \mathbb{C}^n$. We claim that $\mathbb{C}^n - B$ is dense in $\mathbb{C}^n$. In fact, if $\{U_\nu\}$ is a sequence of open sets in Ω such that $p_o | U_\nu$ is a homeomorphism onto an open set in $\mathbb{C}^n$, and K'_ν a compact set in U_ν with $\bigcup K'_\nu = \Omega$, then $B = \bigcup p_o(A \cap K'_\nu)$. Each $p_o(A \cap K'_\nu)$ is a compact nowhere dense set in $\mathbb{C}^n$, so that $\mathbb{C}^n - B = \bigcap(\mathbb{C}^n - p_o(A \cap K'_\nu))$ is dense by Baire's theorem. Let $\{z_\nu\}$ be a dense sequence in $\mathbb{C}^n$, $z_\nu \notin B$. Let $E = \bigcup p_o^{-1}(z_\nu)$. Then E is dense in Ω, $E \cap A = \emptyset$ and $p^{-1}p(E) = E$. Moreover, by the Poincaré-Volterra theorem (chapter 2) E is countable.

Consider now the space $\mathcal{H}(\Omega)$ with the topology of compact convergence [i.e., a sequence in $\mathcal{H}(\Omega)$ converges if it converges uniformly on every compact subset of Ω]. With this topology, $\mathcal{H}(\Omega)$ is a complete metric space (it is in fact a closed subspace of the space of continuous functions with the topology of compact convergence).

Let $H_o = \{f \cdot h | \; f \in \mathcal{H}(\Omega)\}$ where h is the function (2) constructed above, and let H be the closure of H_o in $\mathcal{H}(\Omega)$. [Actually, it can be shown that H_0 is closed, but this is of no importance.] Then clearly any $F \in H$, $F \neq 0$, has zeros of arbitrarily large order in each P_ν.

For any two elements $x, x' \in E$, $x \neq x'$, denote by $H(x, x')$ the set $\{F \in H \mid F(x) \neq F(x')\}$. Clearly, $H(x, x')$ is open in H. We claim that it is dense in H. First, there is $G \in H$ such that $G(x) \neq G(x')$ [for if $h(x) \neq h(x')$, there is nothing to prove. If $h(x) = h(x')$, then, since $E \cap A = \emptyset$, $h(x) \neq 0$, so that if $f \in \mathcal{H}(\Omega)$ separates x and x', then $G = fh \in H$ and separates x, x']. Let $F \in H$, $F \notin H(x, x')$. Then $F + \varepsilon G \in H(x, x')$ for all $\varepsilon \neq 0$, so that $F \in \overline{H(x, x')}$. Hence $H(x, x')$ is dense in H.

Now H, being a closed subspace of a complete metric space, is again a complete metric space. Since E is countable

$$\bigcap_{\substack{x, x' \in E, \\ x \neq x'}} H(x, x') = H'$$

is dense in H, in particular $\neq \emptyset$. If $g \in H'$, then g satisfies conditions (a) and (b) of Part 1 of the proof of Theorem 2.

Proof of Theorem 2. Part 2. We shall now prove that if g satisfies (a) and (b) of Part 1, $S = \{g\}$ and $p: X \to \mathbb{C}^n$, $\varphi: \Omega \to X$ is the S-envelope of holomorphy of $p_0: \Omega \to \mathbb{C}^n$, then φ is an analytic isomorphism. Since, in any case, φ is a local analytic isomorphism, it suffices to prove that φ is bijective.

(i) φ is injective. If we identify X with a connected component of $\mathcal{O} = \mathcal{O}(S)$, we see that φ is injective if and only if the following holds:

If $x, y \in \Omega$, $p_o(x) = p_o(y)$, and U, V small enough connected neighborhoods of x, y respectively such that $p_o(U) = p_o(V) = W$, then $g \circ (p_o|U)^{-1} \not\equiv g \circ (p_o|V)^{-1}$ on W. But if $a \in W$ and $(p_o|U)^{-1}(a) = \xi$ $(p_o|V)^{-1}(a) = \eta$ and $\xi, \eta \in E$, we have $g(\xi) \neq g(\eta)$, so that the above condition is verified. Thus φ is injective.

(ii) <u>φ is surjective.</u> Suppose that φ is not surjective, and let $Y = \varphi(\Omega)$. Then Y is an open set in X, $Y \neq X$. Let $x_o \in \overline{Y} - Y$ (closure being in X), and let P be a polydisc of radius $\rho > 0$ about x_o in X. Let Q be the polydisc of radius $\rho/4$ about x_o, and let $x_\nu \in \Omega$ be so that $\varphi(x_\nu) \in Q \cap Y$ (note that $\{x_\nu\}$ is dense in Ω). Then clearly $\varphi(P_\nu)$ is relatively compact in X. If $G \in \mathcal{H}(X)$ is such that $G \circ \varphi = g$, then the orders of the zeros of G on $\varphi(P_\nu)$ are bounded, contradicting (a) of Part 1. Hence φ is surjective, and the theorem is proved.

<u>Corollary 1</u>. If $p_o : \Omega \to \mathbb{C}^n$ is a domain of holomorphy it is also the domain of existence of an element $g \in \Omega$. [This implies that for any $S \subset \mathcal{H}(\Omega)$, any connected component of $\mathcal{O}(S)$ is analytically isomorphic to a connected component of $\mathcal{O}$.]

This follows from Theorems 1 and 2 and chapter 5, Corollary 2 to Lemma 2.

<u>Corollary 2.</u> If $p_o : \Omega \to \mathbb{C}^n$ is such that $d(\hat{K}) > 0$ for any compact $K \subset \Omega$ and $\mathcal{H}(\Omega)$ separates points of Ω, then $p_o : \Omega \to \mathbb{C}^n$ is a domain of holomorphy and $d(K) = d(\hat{K})$.

Corollary 3. If $\mathcal{H}(\Omega)$ separates points of Ω and $\hat{K}$ is compact for any compact $K \subset \Omega$, then Ω is a domain of holomorphy. In particular, $\Omega \subset \mathbb{C}^n$ is a domain of holomorphy if and only if $\hat{K}$ is compact for any compact set $K \subset \Omega$.

Definition 6. Given $p_0: \Omega \to \mathbb{C}^n$ and $S \subset \mathcal{H}(\Omega)$, we say that Ω is S-convex if, for every compact set $K \subset \Omega$, the set $\hat{K}_S$ is again compact.

Proposition 2. Let $p_0: \Omega \to \mathbb{C}^n$ be a domain of holomorphy and let K be a compact set in Ω. For any $A \subset \Omega$ with $d(A) > 0$, set $A(r) = \bigcup_{x \in A} \overline{P}(x, r)$, $0 < r < d(A)$; here $\overline{P}(x, r)$ is the closure in Ω of the polydisc of radius r about x. Then, for $0 < r < d(K)$, $L = K(r)$ is compact and we have

$$\hat{K}(r) \subset \hat{L}.$$

Proof. Note that $\hat{K}(r)$ is defined since $d(K) = d(\hat{K})$. Suppose the inclusion false, and let $x_0 \in \hat{K}(r)$, $x_0 \notin \hat{L}$. Then, there is $f \in \mathcal{H}(\Omega)$ so that

$$f(x_0) = 1 \ , \ \|f\|_{K(r)} < 1 \ .$$

The function $g = \frac{1}{1-f}$ is holomorphic in a neighborhood U of $\hat{L}$. Let $\rho > 0$, $r < \rho < d(K)$ be such that the closure of the polydisc $P(a,\rho)$ of radius ρ about $a \in K$ is contained in U. It follows from Cauchy's inequalities that

$$|D^\alpha g(a)| \leq M \rho^{-|\alpha|} . \alpha! \quad , \quad M = \|g\|_{K(\rho)} ,$$

so that

$$\|D^\alpha g\|_K \leq M\rho^{-|\alpha|} . \alpha! \quad .$$

Now, $g = \lim_{N\to\infty} g_N$, $g_N = \sum_{p=0}^{N} f^p \in \mathcal{H}(\Omega)$, and the limit is uniform on a neighborhood of $\hat{L}$ (since $\|f\|_{K(r)} < 1$). Hence $D^\alpha g_N$ converges uniformly to $D^\alpha g$ on a neighborhood of $\hat{L}$. In particular, for $b \in \hat{K}$, we have

$$|D^\alpha g(b)| = \lim_{N\to\infty} |D^\alpha g_N(b)| \leq \lim_{N\to\infty} \|D^\alpha g_N\|_K = \|D^\alpha g\|_K ,$$

so that

$$\|D^\alpha g\|_{\hat{K}} \leq M\rho^{-|\alpha|} . \alpha! \ .$$

It follows at once that the Taylor series of g,

$$\sum_{\alpha \in \mathbb{N}^m} \frac{D^\alpha g(b)}{\rho!} (p_0(x) - p_0(b))^\alpha ,$$

converges uniformly to g on the polydisc $P(b, \rho')$ for any $b \in \hat{K}$ and $r < \rho' < \rho$. In particular, if $b \in \hat{K}$ and $\rho' > r$ are so chosen that $x_0 \in P(b, \rho')$ (note that $x_0 \in \hat{K}(r)$), the function g can be continued holomorphically to $P(b, \rho')$, hence to a neighborhood of x_0, which is absurd. This contradiction shows that $\hat{K}(r) \subset \hat{L}$.

Corollary. If $p_0 : \Omega \to \mathbb{C}^n$ is a domain of holomorphy and K_0, K_1 are compact sets with $K_o \subset \overset{o}{K}_1$, then, we have

$$\hat{K}_0 \subset (\hat{K}_1)^o .$$

Proof. Choose r so small that $K_0(r) \subset K_1$. Then, by Proposition 2,

$$\hat{K}_0(r) \subset \hat{K}_1 .$$

Since clearly $\hat{K}_0$ is contained in the interior of $\hat{K}_0(r)$, the result follows.

Proposition 3. Let $p_o: \Omega \to \mathbb{C}^n$ be a connected domain over $\mathbb{C}^n$ and $p: X \to \mathbb{C}^n$, $\varphi: \Omega \to X$ its envelope of holomorphy. Let $\{K_p\}$ be a sequence of compact sets in Ω such that $K_p \subset \overset{o}{K}_{p+1}$, $\bigcup K_p = \Omega$. Let $L_p = \varphi(K_p)$, and $Q_p = \hat{L}_p$ (relative to X). Then

$$Q_p \subset \overset{o}{Q}_{p+1} \quad \text{and} \quad \bigcup Q_p = X.$$

Proof. Since φ is open, we have $L_p \subset \overset{o}{L}_{p+1}$. Moreover, by chapter 6, Corollary to Lemma 1, $p: X \to \mathbb{C}^n$ is a domain of holomorphy. Hence, by the corollary to Proposition 2 above, $Q_p \subset \overset{o}{Q}_{p+1}$. In particular, $Y = \bigcup Q_p$ is open in X. Suppose that $Y \neq X$. Let $x_o \in \overline{Y} - Y$, and let $\{y_\nu\}$ be a sequence in Y, $y_\nu \to x_o$. Now, any compact set $L \subset Y$ is contained in Q_p for some p. Since $\lim_{\nu \to \infty} y_\nu = x_o \notin \bigcup Q_p$, it follows that no Q_p contains all the y . Moreover, if $L \subset Q_p$, then $\hat{L}$ (relative to Y) is also contained in Q_p (note that any holomorphic function g on Y can be continued analytically to X; it is sufficient to find $G \in \mathcal{H}(X)$ so that $G \circ \varphi = g \circ \varphi$). Hence, by Lemma 2, there exists $g \in \mathcal{H}(Y)$ such that $\{g(y_\nu)\}$ is not bounded. If $G \in \mathcal{H}(X)$ is such that $G|Y = g$, then $\{G(y_\nu)\}$ is not bounded. This is absurd since $y_\nu \to x_o$. This contradiction shows that $\bigcup Q_p = X$.

Corollary 1. Let $p_o: \Omega \to \mathbb{C}^n$, $p: X \to \mathbb{C}^n$, $\varphi: \Omega \to X$ be as above. Then, for any compact set $K \subset X$, there exists a compact set $L \subset \Omega$ such that $K \subset \hat{Q}$, where $Q = \varphi(L)$.

Corollary 2. Let $\{f_\nu\}$ be a sequence of holomorphic functions on Ω and let $F_\nu \in \mathcal{H}(X)$ be such that $F_\nu \circ \varphi = f_\nu$. If $\{f_\nu\}$ converges uniformly on compact subsets of Ω, then $\{F_\nu\}$ converges uniformly on compact subsets of X.

Corollary 3. The map $\mathcal{H}(X) \to \mathcal{H}(\Omega)$, $F \mapsto F \circ \varphi$, is a topological isomorphism of these $\mathbb{C}$-algebras, provided with the topology of uniform convergence on compact sets.

We shall now give some applications of these results.

Let $p_0 : \Omega \to \mathbb{C}^n$ and $p'_0 : \Omega' \to \mathbb{C}^{n'}$ be connected domains over $\mathbb{C}^n$, $\mathbb{C}^{n'}$ respectively and let $p : X \to \mathbb{C}^n$, $\varphi : \Omega \to X$ and $p' : X' \to \mathbb{C}^{n'}$, $\varphi' : \Omega' \to X'$ be their envelopes of holomorphy. The map

$$q_0 : \Omega \times \Omega' \to \mathbb{C}^{n+n'}, \quad q_0 = p_0 \times p'_0 : (x, x') \mapsto (p_0(x), p'_0(x'))$$

is a connected domain over $\mathbb{C}^{n+n'}$. Similarly, $q = p \times p' : X \times X' \to \mathbb{C}^{n+n'}$ is a domain over $\mathbb{C}^{n+n'}$ and $\psi = \varphi \times \varphi' : \Omega \times \Omega' \to X \times X'$ is a local isomorphism.

Proposition 4. $q : X \times X' \to \mathbb{C}^{n+n'}$, $\psi : \Omega \times \Omega' \to X \times X'$ is the envelope of holomorphy of $q_0 : \Omega \times \Omega' \to \mathbb{C}^{n+n'}$.

Proof. Let K, K' be compact sets in X, X' respectively, and let $L = K \times K'$. We claim that $\hat{L} \subset \hat{K} \times \hat{K}'$. In fact, if $x \notin \hat{K}$, there is f in $\mathcal{H}(X)$ so that $|f(x)| \geq 1$, $\|f\|_K < 1$. Hence the function $F \in \mathcal{H}(X \times X')$ defined by $F(y, y') = f(y)$ has the property

$$|F(x, x')| \geq 1 \quad , \quad \|F\|_{K \times K'} < 1$$

so that $(x, x') \notin \hat{L}$. A similar reasoning applies if $x' \notin \hat{K}'$. We have, obviously, $d_q(\hat{K} \times \hat{K}') \geq \min(d_p(\hat{K}), d_{p'}(\hat{K}')) > 0$ (by Theorem 1). Moreover, $\mathcal{H}(X \times X')$ clearly separates the points of $X \times X'$. Hence, by Corollary 2 to Theorem 2, $q : X \times X' \to \mathbb{C}^{n+n'}$ is a domain of holomorphy. To complete the proof of the theorem it suffices to show that $X \times X'$ is an $\mathcal{H}(\Omega \times \Omega')$-extension of $q_0 : \Omega \times \Omega' \to \mathbb{C}^{n+n'}$. Let

P be a polydisc about $a \in \Omega$. Then, for any polydisc $P' \subset \Omega'$, $f|P \times P'$ can be expanded in a series

$$f(x,x') = \sum_{\alpha \in \mathbb{N}^n} (p_o(x) - p_o(a))^\alpha f'_\alpha(x') \ , \qquad (x,x') \in P \times P' .$$

Moreover, the functions $f_\alpha(x')$ are uniquely determined. It follows that

$$f(x,x') = \sum_{\alpha \in \mathbb{N}^n} (p_o(x) - p_o(a))^\alpha f'_\alpha(x') \ , \quad f_\alpha \in \mathcal{H}(\Omega')$$

and the series converges uniformly on compact subsets of $P \times \Omega'$. Let $K' \subset \Omega'$ be compact, let $L' = \varphi'(K')$ and $Q' = \hat{L}'$ in X'. Let $g'_\alpha \in \mathcal{H}(X')$ be so that $g'_\alpha \circ \varphi' = f'_\alpha$. Now

$$\|f'_\alpha\|_{K'} \leq \text{const.}\, \rho^{-|\alpha|} \qquad \text{for} \quad 0 < \rho < \text{radius}\ (P).$$

Hence, $\|g'_\alpha\|_{Q'} \leq \|f'_\alpha\|_{K'} \leq \text{const.}\, \rho^{-|\alpha|}$. Hence the series

$$g_P(x,x') = \sum_{\alpha \in \mathbb{N}^n} (p_o(x) - p_o(a))^\alpha g'_\alpha(x') \ , \qquad (x,x') \in P \times X'$$

converges (by Corollary 1 to Proposition 3) uniformly on compact subsets of $P \times X'$. Hence, there is $g_P \in \mathcal{H}(P \times X')$ such that

$$g_P(x, \varphi'(x')) = f(x,x') \ , \quad (x,x') \in P \times \Omega' .$$

By the uniqueness of the extension, it follows that there is $g \in \mathcal{H}(\Omega \times X')$ so that $g(x, \varphi'(x')) = f(x,x')$, $(x,x') \in \Omega \times \Omega'$. Repeating this argument with Ω replaced by X' and Ω' by Ω, we find that there is $h \in \mathcal{H}(X \times X')$ so that

$$h(\varphi(x), x') = g(x,x') \quad , \ (x,x') \in \Omega \times X'.$$

Clearly, $h \circ \psi = f$, so that $q: X \times X' \to \mathbb{C}^{n+n'}$, $\psi: \Omega \times \Omega' \to X \times X'$ is an $\mathcal{H}(\Omega \times \Omega')$-extension of $q_o: \Omega \times \Omega' \to \mathbb{C}^{n+n'}$, and the result follows.

Let $\Omega \subset \mathbb{C}^n$ be a Reinhardt domain, $0 \in \Omega$. Let $B \subset \mathbb{R}^n$ be the set

$$B = \{(x_1, \ldots, x_n) \in \mathbb{R}^n \mid (e^{x_1}, \ldots, e^{x_n}) \in \Omega\}.$$

We say that Ω is logarithmically convex if B is convex in $\mathbb{R}^n$. We say that it is complete if $z \in \Omega$, $z' \in \mathbb{C}^n$ and $|z_j'| \leq |z_j|$, $j = 1, \ldots, n$ implies $z' \in \Omega$.

Proposition 5. A Reinhardt domain containing 0 is a domain of holomorphy if and only if it is logarithmically convex and complete. The envelope of holomorphy of a Reinhardt domain is the smallest logarithmically convex complete Reinhardt domain containing it.

Proof. It is sufficient to prove the first part of the proposition.

Let Ω be a Reinhardt domain containing 0 which is a domain of holomorphy. Let $f \in \mathcal{H}(\Omega)$ and let $f(z) = \sum_{\alpha \in \mathbb{N}^n} a_\alpha z^\alpha$ be its Taylor expansion about 0. Let $\Omega_f = \{a \in \mathbb{C}^n \mid \sum a_\alpha z^\alpha$ converges uniformly in the neigborhood of $a\}$. Then

$$\Omega = \bigcap_{f \in \mathcal{H}(\Omega)} \Omega_f .$$

Clearly, if $z \in \Omega$ and $z' \in \mathbb{C}^n$ is such that $|z_j'| \leq |z_j|$, then $z' \in \Omega_f$ for all f. Hence Ω is complete. Let $B_f = \{x \in \mathbb{R}^n \mid (e^{x_1}, \ldots, e^{x_n}) \in \Omega_f\}$. It is sufficient to prove that B_f is convex for any $f \in \mathcal{H}(\Omega)$. Now, B_f is the interior of the set A_f of $x \in \mathbb{R}^n$ such that there exists $M(x) > 0$ such that

$$|a_\alpha| e^{\alpha_1 x_1 + \ldots + \alpha_n x_n} \leq M(x) \quad \text{for all } \alpha \in \mathbb{N}^n .$$

Hence

$$A_f = \{x \in \mathbb{R}^n \mid \exists M(x) > 0 \text{ with } \alpha_1 x_1 + \ldots + \alpha_n x_n \leq \log M(x) - \log|a_\alpha| \\ \text{for all } \alpha \text{ such that } a_\alpha \neq 0\}.$$

This latter set is obviously convex, hence so is B_f.

To prove the converse we proceed as follows. Let K be compact in Ω. Then, there is a finite set $S \subset \Omega$ so that

$$K \subset \bigcup_{a \in S} \{z \in \mathbb{C}^n \mid |z_j| \leq |a_j|, \ j = 1, \ldots, n\}.$$

We may clearly suppose that none of the $|a_j|$ is 0 for $a \in S$. Let L be the closure in $\mathbb{C}^n$ of $\hat{K}$. It suffices to prove that $L \subset \Omega$. Let $z \in L$, $z = (z_1, \ldots, z_n)$. Let $i_1, \ldots, i_k$ be those indices i with $z_i \neq 0$. Then, for any integers $\alpha_1, \ldots, \alpha_k \geq 0$, we have

$$|z_{i_1}^{\alpha_1} \cdots z_{i_k}^{\alpha_k}| \leq \sup_{w \in K} |w_{i_1}^{\alpha_1} \cdots w_{i_k}^{\alpha_k}| \leq \sup_{a \in S} |a_{i_1}^{\alpha_1} \cdots a_{i_k}^{\alpha_k}|,$$

i.e.,

$$\sum_{\nu=1}^{k} \alpha_\nu \log |z_{i_\nu}| \leq \sup_{a \in S} \sum_{\nu=1}^{k} \alpha_\nu \log |a_{i_\nu}| .$$

Since this holds for all integers $\alpha_\nu \geq 0$, it holds for all rational $\alpha_\nu \geq 0$, hence for all real $\alpha_\nu \geq 0$. Hence $(\log|z_{i_1}|, \ldots, \log|a_{i_k}|)$ is in the convex envelope of the points $(\log|a_{i_1}|, \ldots, \log|a_{i_k}|)$, $a \in S$. Since the projection on the $(x_{i_1}, \ldots, x_{i_k})$-plane of the convex set $B = \{(x_1, \ldots, x_n) \in \mathbb{R}^n \mid (e^{x_1}, \ldots, e^{x_n}) \in \Omega\}$ is again convex, there is a point $\zeta = (\zeta_1, \ldots, \zeta_n) \in \Omega$ such that

$$\log|z_{i_\nu}| = \log|\zeta_{i_\nu}|, \quad \nu = 1, \ldots, k .$$

Clearly $|z_j| \leq |\zeta_j|$ for all $j = 1, \ldots, n$. Since $\zeta \in \Omega$ and Ω is complete, we have $z \in \Omega$, so that $L \subset \Omega$.

Let $B \subset \mathbb{R}^n$. By the <u>tube</u> T_B on B we mean the set $\{z \in \mathbb{C}^n \mid (\mathrm{Re}\, z_1, \ldots, \mathrm{Re}\, z_n) \in B\}$.

<u>Proposition 6</u>. Let B be a connected open set in $\mathbb{R}^n$, $n \geq 2$, and T_B the tube on B. The envelope of holomorphy of T_B is $T_{\hat{B}}$, where $\hat{B}$ is the convex envelope of B in $\mathbb{R}^n$.

We remark first the following.

<u>Lemma 8</u>. Any convex open set $\Omega \subset \mathbb{C}^n$ is a domain of holomorphy.

<u>Proof</u>. Let $a \in \partial\Omega$. Then, since Ω is convex, there is a function

$$\ell(z) = \lambda_1 z_1 + \ldots + \lambda_n z_n + \lambda,$$

$\lambda, \lambda_j \in \mathbb{C}$ such that $\ell(a) = 0$ and $\mathrm{Re}\, \ell(z) < 0$ for $z \in \Omega$. If K is a compact set in Ω, there is therefore a neighborhood U_a of a such that, if we set $\varphi(z) = \exp(\ell(z))$, then $\|\varphi\|_K < |\varphi(z)|$ for any $z \in U_a$. In particular, $\hat{K} \cap U_a = \emptyset$. Since $a \in \partial\Omega$ is arbitrary, it follows that $\hat{K}$ is closed in $\mathbb{C}^n$, hence is compact. Hence Ω is a domain of holomorphy.

<u>Corollary.</u> If $B \subset \mathbb{R}^n$ is a convex open set, then T_B is a domain of holomorphy.

In view of this corollary, to prove Proposition 6, it is sufficient to show that if B is connected, any $f \in \mathcal{H}(T_B)$ can be extended to $T_{\hat{B}}$.

<u>Lemma 9</u>. Let $a_o = (1, 0, \ldots, 0)$ and $a_1 = (0, 1, 0, \ldots, 0) \in \mathbb{R}^n$, $n \geq 2$. Let A be the triangle $\{(x_1, x_2, 0, \ldots, 0),\ x_1 \geq 0,\ x_2 \geq 0,$ $x_1 + x_2 \leq 1\}$ and, for $0 \leq \lambda < 1$, let A_λ be the triangle $\{(x_1, x_2, 0, \ldots, 0),$

$x_1 \geq 0,\ x_2 \geq 0,\ x_1 + x_2 \leq \lambda\}$. Let B be an open set in $\mathbb{R}^n$ containing A, and $\Omega = T_B$. For any λ, $0 \leq \lambda < 1$, there is $M = M_\lambda > 0$ such that if $K = \{z \in \mathbb{C}^n \mid \operatorname{Re} z \in \Gamma,\ |\operatorname{Im} z| \leq M\}$, then $\hat{K}_{\mathcal{H}(\Omega)}$ contains A_λ; here $\Gamma = \{ta_o \mid 0 \leq t \leq 1\} \cup \{ta_1 \mid 0 \leq t \leq 1\}$.

Proof. Let $\varepsilon > 0$ be sufficiently small, and let

$$S'_\varepsilon = \{z \in \mathbb{C}^n \mid z_1 + z_2 - \varepsilon(z_1^2 + z_2^2) = 1 - \varepsilon\ ,\ z_3 = \ldots = z_n = 0\}.$$

and $S_\varepsilon = S'_\varepsilon \cap T_A$. We claim that for any $f \in \mathcal{H}(\Omega)$, we have

$$\|f\|_{S_\varepsilon} = \|f\|_{S_\varepsilon \cap \Gamma'}\ ,\quad \Gamma' = T_\Gamma .$$

If we write $z_j = x_j + iy_j$, we have, on S_ε,

$$x_1 + x_2 - \varepsilon(x_1^2 + x_2^2) + \varepsilon(y_1^2 + y_2^2) = 1 - \varepsilon\ ,\quad x_1 \geq 0,\ x_2 \geq 0,\ x_1 + x_2 \leq 1.$$

In particular, $y_1^2 + y_2^2 \leq \frac{1-\varepsilon}{\varepsilon}$. Since, moreover $0 \leq x_1 \leq 1$, $0 \leq x_2 \leq 1$, it follows that S_ε is compact. Furthermore, $x_1 + x_2 < 1$ on S_ε except at the points a_o and a_1.

Suppose now that $|f|S_\varepsilon|$ attains a maximum at a point $\zeta \in S_\varepsilon - \Gamma'$. If $\zeta = (\zeta_o, \zeta_1, 0, \ldots, 0)$, there is a holomorphic function $\varphi(u)$ in a connected neighborhood U of $u = \zeta_o$ in $\mathbb{C}$ such that $\varphi(\zeta_o) = \zeta_1$ and the set $\{(u, \varphi(u), 0, \ldots, 0)\}$ is a neighborhood of ζ on S_ε. Hence $|f(u, \varphi(u), 0, \ldots, 0)|$ has a maximum at $u = \zeta_o$ and so is constant on U. It follows that the set of $z \in S_\varepsilon - \Gamma'$ at which $|f|S_\varepsilon|$ has a maximum is open and it is obviously also closed. Hence $\|f\|_{S_\varepsilon} = \|f\|_{S_\varepsilon \cap \Gamma'}$.

It is easily verified that if $\lambda < 1$, and ε is small enough, for any $x \in A_\lambda$, there is $y = (y_1, y_2, 0, \ldots, 0)$ such that $x \in S_\varepsilon + iy$, $|y| \leq \frac{1}{\varepsilon}$. Hence, $|f(x)| \leq \|f\|_{(S_\varepsilon + iy) \cap \Gamma'}$. If we set

$$M = 2/\varepsilon\ ,\quad K = \{z \in \mathbb{C}^n \mid \operatorname{Re} z \in \Gamma,\ |\operatorname{Im} z| \leq M\},$$

this implies that

$$\|f\|_{A_\lambda} \leq \|f\|_K ,$$

so that $A_\lambda \subset \hat{K}$.

Lemma 10. Let $a_0, a_1, \Gamma, A_\lambda$, A be as in Lemma 9. Let B be the union of two open convex sets in $\mathbb{R}^n$ containing, respectively, the sets $\{ta_0 \mid 0 \leq t \leq 1\}$ and $\{ta_1 \mid 0 \leq t \leq 1\}$. Let $\Omega = T_B$. Then any $f \in \mathcal{H}(\Omega)$ can be extended holomorphically to a neighborhood of T_A.

Proof. Let E be the set of λ, $0 \leq \lambda \leq 1$, such that there is a convex neighborhood U of A_λ in $\mathbb{R}^n$ with the property that any $f \in \mathcal{H}(\Omega)$ can be extended to T_U. (Note that $T_U \cap T_B$ is connected.) E is clearly open in $[0,1]$ and $0 \in E$. Let $\lambda \in \overline{E}$.

Let $\rho = d_\Omega(\Gamma)$ be the distance of Γ from the boundary of Ω. Clearly $\rho > 0$. Let $\mu \in E$, $|\lambda - \mu| < r < \rho$ and let U be a convex neighborhood of A_μ such that any $f \in \mathcal{H}(\Omega)$ can be extended to T_U. Then $B' = U \cup B$ is connected, and so is $U \cap B$. Any $f \in \mathcal{H}(\Omega)$ can therefore be extended to $T_{B'} = \Omega'$. Also $d_{\Omega'}(\Gamma) \geq \rho$. Let $\mu_0 < \mu$, $|\lambda - \mu_0| < r < \rho$. By Lemma 9, there is $M > 0$ such that if $K = \{z \in \mathbb{C}^n \mid \operatorname{Re} z = \Gamma, \ |\operatorname{Im} z| \leq M\}$, then $\hat{K}$ (relative to Ω') contains A_{μ_0}. By Proposition 1, the Taylor series of any $f \in \mathcal{H}(\Omega')$ about any point $a \in A_{\mu_0}$ converges in the polydisc of radius ρ about a. By con- considering the functions $f_y: z \mapsto f(z + iy)$, $y \in \mathbb{R}^n$, one sees that the same is therefore true for any $a \in T_{A_{\mu_0}}$. It follows that there is a convex neighborhood V of A_λ such that any $f \in \mathcal{H}(\Omega')$ can be extended to T_V. Hence $\lambda \in E$, so that E is closed. Hence $E = [0,1]$ and the lemma is proved.

Proof of Proposition 6. As remarked earlier, it suffices to show that any $f \in \mathcal{H}(\Omega)$, $\Omega = T_B$, can be extended to $T_{\hat{B}}$. For this, we first show that if $a, b \in B$, and $\ell(a,b)$ denotes the closed line segment joining a to b, there is a convex neighborhood U of $\ell(a,b)$ and convex neighborhoods U_a of a, U_b of b such that for any $f \in \mathcal{H}(\Omega)$, there is $F \in \mathcal{H}(T_U)$ which coincides with f on T_{U_a}, T_{U_b}. If this property holds, we say that B can be extended to $\ell(a,b)$.

For any $x_0, x_1, x_2 \in B$, denote by $A(x_0, x_1, x_2)$, the closed triangle in $\mathbb{R}^n$ with vertices x_0, x_1, x_2. From Lemma 9, it follows that if B extends to $\ell(x_0, x_1)$ and to $\ell(x_0, x_2)$, then B extends to $\ell(x_1, x_2)$.

Given $a, b \in B$, let x_0 be a fixed point in B and choose two polygons (with vertices $a_0 = x_0$, $a_1, \ldots, a_p = a$ and $b_0 = x_0$, $b_1, \ldots, b_p = b$ respectively) joining x_0 to a and to b respectively. By our remark above, B extends to $\ell(a_1, b_1)$. Since B obviously extends to $\ell(a_1, a_2)$, B extends to $\ell(b_1, a_2)$. Since B obviously extends to $\ell(b_1, b_2)$, it follows that B extends to $\ell(a_2, b_2)$. Continuing in this way, we see that B extends to $\ell(a_p, b_p) = \ell(a,b)$, and the result is proved.

To complete the proof of the theorem, it suffices now to show that if $x \in \hat{B}$ and x lies on $\ell(a,b)$ and on $\ell(a', b')$, $a, b, a', b' \in B$, and $f \in \mathcal{H}(T_B)$, the functions F and F' obtained on T_U (U a convex neighborhood of x) by extending B to $\ell(a,b)$ and to $\ell(a', b')$ coincide. It follows from our remark above that there is G holomorphic on T_W (where W is a convex neighborhood of $A(a, a', x)$) such that $G = f$ in a neighborhood of $T_{\{a\}} \cup T_{\{a'\}}$ and $G = F$ in a neighborhood of $T_{\{x\}}$.

Similarly, there is $G' \in \mathcal{H}(T_W)$ with $G' = f$ in a neighborhood of $T_{\{a\}} \cup T_{\{a'\}}$ and $G' = F'$ in a neighborhood of $T_{\{x\}}$. But then $G = G'$ on T_W (since $G = G'$ in a nonempty open subset of T_W), so that $F = F'$.

This proves Proposition 6.

Remarks. 1. It can be proved that if B is a connected open set in $\mathbb{R}^n$ and $\Omega = T_B$, we have the following.

For any compact set $K \subset \Omega$, $\varepsilon > 0$ and $f \in \mathcal{H}(\Omega)$, there exist linear functions

$$\ell_\nu(z) = \sum_{j=1}^{n} \lambda_{\nu,j} z_j \,, \quad \lambda_{\nu,j} \in \mathbb{R} \,,$$

and constants $\alpha_\nu \in \mathbb{C}$, $\nu = 1, \ldots, p$, such that

$$\left| f(z) - \sum_{\nu=1}^{p} \alpha_\nu e^{\ell_\nu(z)} \right| < \varepsilon \qquad \text{for all } z \in K.$$

This can be looked upon as the analogue of the expansion in a power series valid for a Reinhardt domain.

2. There is a close relation between Reinhardt domains and tubes. Proposition 5 corresponds essentially to the special case of Proposition 6 which asserts that all $f \in \mathcal{H}(T_B)$ which are periodic (i.e., for which there is $y \in \mathbb{R}^n$, $y = (y_1, \ldots, y_n)$, $y_j > 0$, such that $f(z) = f(z + iy)$) extend to $\mathcal{H}(T_B)$.

3. The above domains are examples of domains Ω which are S-convex for very small families S. Proposition 5 asserts that a Reinhardt domain Ω containing 0 which is $\mathcal{H}(\Omega)$-convex is S-convex where S is the family of monomials $z \mapsto z^\alpha$, $\alpha \in \mathbb{N}^n$. Proposition 6 asserts that a tube $\Omega = T_B$ which is $\mathcal{H}(\Omega)$-convex is S-convex where S

is the family of linear functions $z \mapsto \lambda_1 z_1 + \ldots + \lambda_n z_n$, $\lambda_j \in \mathbb{C}$. In particular, these domains are convex with respect to the family of functions holomorphic on $\mathbb{C}^n$.

We end this chapter with a sufficient condition for a domain in $\mathbb{C}^n$ to be a domain of holomorphy.

Proposition 7. Let D be a bounded domain in $\mathbb{C}^n$. Suppose that there is a compact set K such that for any $x \in D$ there is an analytic automorphism $\sigma \in \mathrm{Aut}(D)$ and a point $a \in K$ such that $\sigma(x) = a$. Then D is a domain of holomorphy.

Proof. Let $p: X \to \mathbb{C}^n$, $\varphi: D \to X$ be the envelope of holomorphy of D. Then φ is injective and to prove the proposition, we have to show that φ is surjective. Suppose this were false. Let $\{x_\nu\}$ be a sequence of points of D such that $\varphi(x_\nu)$ converges to a point $y_0 \in \partial\varphi(D)$. Let $a_\nu \in K$ and $\sigma_\nu \in \mathrm{Aut}(D)$ be such that $\sigma_\nu(x_\nu) = a_\nu$. Let P be a polydisc about y_0 in X which is relatively compact in X. By chapter 6, Corollary to Proposition 3, there is an automorphism $\tilde{\sigma}_\nu$ of X such that $\tilde{\sigma}_\nu \circ \varphi = \sigma_\nu$. Further, since D is bounded, chapter 6, Proposition 2 implies that $p \circ \tilde{\sigma}_\nu$ is bounded, uniformly with respect to ν. Hence, from Cauchy's inequality and the mean value theorem, we obtain the following:

Let $P_\rho \subset P$ be the polydisc of radius ρ about y_0. There is a constant $C > 0$ (independent of ρ) such that if $y \in P_\rho$, we have $|\tilde{\sigma}_\nu(x) - \tilde{\sigma}_\nu(y)| \leq C\rho$ for all $x \in P_\rho$. Let ρ be sufficiently small. We obtain the following: There is a compact set $L \subset D$ such that

$$\sigma_\nu(\varphi^{-1}(P_\rho \cap \varphi(D)) = \sigma_\nu(\varphi^{-1}(P_\rho)) \subset L.$$

Choose now a subsequence $\{\nu_p\} \subset \{\nu\}$ such that $\{\sigma_{\nu_p}\}$ and $\{\sigma_{\nu_p}^{-1}\}$ converge uniformly on compact subsets of D to mappings $\sigma, \sigma' : D \to \mathbb{C}^n$. Then $\sigma(\varphi^{-1}(P_p)) \subset L$. Hence, by chapter 5, Theorem 4, $\sigma \in \mathrm{Aut}(D)$ and σ' is its inverse, $\sigma' \circ \sigma = \sigma \circ \sigma' =$ identity. But this is absurd since $\sigma_\nu^{-1}(a_\nu) = x_\nu$, if a is a limit point of $\{a_{\nu_p}\}$ in K, $\sigma'(a) \in D$. But $\{x_\nu\}$ has no limit point in D. This contradiction proves the proposition.

<u>Corollary 1</u>. If Γ is a discrete subgroup of Aut(D) such that D/Γ is compact, then D is a domain of holomorphy.

<u>Corollary 2.</u> If D is a bounded homogeneous domain, i.e., for any pair of points $x, y \in D$, there is $\sigma \in \mathrm{Aut}(D)$ such that $\sigma(x) = y$, then D is a domain of holomorphy.

There is an outstanding conjecture that if D is a bounded domain in $\mathbb{C}^n$ and if there exists a discrete group $\Gamma \subset \mathrm{Aut}(D)$ such that D/Γ is compact, then D <u>is homogeneous</u>. Nothing is known about this question.

For the results of this chapter, see [1], [10], [19], [11].

The results on Reinhardt domains were inspired by [27]. For the form given here, see [17], [19].

The theorem on tubes is due to Bochner [3]. See also [17].

Corollary 1 of Proposition 7 above is due to M. Hervé (Annales de l'Ecole Normale Sup., 69(1952), 277-302). There is a proof in the notes of C. L. Siegel (Analytic functions of several complex variables, Princeton 1948/49).

Corollary 2 of Proposition 7 above is due to P. Thullen (Math. Annalen, 104 (1931), 373-376).

J. Vey (Ann. Scient. Ec. Norm. Sup. 1970) has proved the conjecture referred to above for a wide class of domains, the so-called Siegel domains; thus his theorem is that if D is a Siegel domain and Γ a discrete subgroup of Aut(D) such that D/Γ is compact, then D is homogeneous. He has also proved an analogue of this result for convex sets in $\mathbb{R}^n$ and discrete groups of affine transformations of D; see Ann. scient. Ec. Norm. Sup., 4^e serie, t.3, 1970, 479-506 and Annali della Scuola Normale Superiore di Pisa, Vol. 24 1970, 641-665.

8

DOMAINS OF HOLOMORPHY: OKA'S THEOREM

We have seen, in chapter 7, that a domain Ω in $\mathbb{C}^n$ is a domain of holomorphy if and only if Ω is $\mathcal{H}(\Omega)$-convex, i.e., if and only if, for any compact set $K \subset \Omega$, the set $\hat{K}$ is again compact. A remarkable theorem, due to K. Oka, implies that the same result is true for domains $p_o: \Omega \to \mathbb{C}^n$ over $\mathbb{C}^n$. We shall give, in this chapter, the proof of one part of this theorem, namely that a domain of holomorphy is $\mathcal{H}(\Omega)$-convex. The converse amounts to showing (in view of the results of chapter 7) that if $p_o: \Omega \to \mathbb{C}^n$ is $\mathcal{H}(\Omega)$-convex, then $\mathcal{H}(\Omega)$ separates the points of Ω. All known proofs of this fact use global ideal theory (Theorems A and B of Oka-Cartan-Serre) and we shall not go into it.

The proof given in this chapter is essentially that of E. Bishop.

Proposition 1 (Hadamard's three domains theorem). Let $p_o: \Omega \to \mathbb{C}^n$ be a connected domain over $\mathbb{C}^n$ and let Ω_o, Ω_1 be nonempty open subsets of Ω with

$$\Omega_o \subset\subset \Omega_1 \subset\subset \Omega.$$

Then, there is α, $0 < \alpha < 1$, such that, for any $f \in \mathcal{H}(\Omega)$, we have

$$\|f\|_{\Omega_1} \leq \|f\|_{\Omega_o}^{\alpha} \|f\|_{\Omega}^{1-\alpha}.$$

Proof. We shall first prove the following result.

(a) For $\rho > 0$, set $B(\rho) = \{z \in \mathbb{C}^n \mid \|z\|^2 = |z_1|^2 + \dots + |z_n|^2 < \rho^2\}$. Let $0 < r_o < r_1 < r$. Then, for $f \in \mathcal{H}(B(r))$, we have

$$\|f\|_{B(r_1)} \le \|f\|^{\alpha}_{B(r_o)} \|f\|^{1-\alpha}_{B(r)} \quad \text{where } \alpha = \frac{\log r - \log r_1}{\log r - \log r_o} .$$

In fact, if $z \in B(r_1)$, $z \neq 0$, chapter 3, Proposition 6, applied to the function $g(u) = f(uz)$ and the three circles of radius $r_o/\|z\|$, $r_1/\|z\|$, $r/\|z\|$ in the u-plane gives us

$$|f(z)| = |g(1)| \le \|f\|^{\alpha}_{B(r_o)} \|f\|^{1-\alpha}_{B(r)} .$$

Let $p_o : \Omega \to \mathbb{C}^n$ be given. An open set $B \subset \Omega$ is called a ball with center $a \in B$ and radius ρ if $p_o | B$ is an isomorphism onto the set

$$\{z \in \mathbb{C}^n \mid \|z - p_o(a)\| < \rho\}.$$

Let $B_o, \dots, B_n$ be a sequence of balls in Ω such that $B_o \cap \Omega_o \neq \emptyset$, the center of B_k is contained in B_{k-1} $(k = 1, \dots, m)$, $B_k \subset\subset \Omega$ and $\bigcup_{k=1}^{m} B_k \supset \Omega_1$. (Such a sequence exists since Ω is connected.)

Set $V_k = B_o \cup \dots \cup B_k$, and let W_k be a ball whose center is that of B_k and such that $W_o \subset\subset \Omega_o \cap B_o$, $W_k \subset\subset V_{k-1} \cap B_k$, $k = 1, \dots, m$. By (a) above, there is α_o, $0 < \alpha_o < 1$, such that

$$\|f\|_{B_o} \le \|f\|^{\alpha_o}_{W_o} \|f\|^{1-\alpha_o}_{\Omega} \quad \text{for all } f \in \mathcal{H}(\Omega).$$

We shall prove, by induction, the existence of α_k, $0 < \alpha_k < 1$ such that

$$(*) \qquad \|f\|_{V_k} \le \|f\|^{\alpha_k}_{W_o} \|f\|^{1-\alpha_k}_{\Omega} .$$

Suppose this proved for a certain k. By (a) above, there is β, $0<\beta<1$ with

$$\|f\|_{B_{k+1}} \le \|f\|^{\beta}_{W_{k+1}} \|f\|^{1-\beta}_{\Omega} .$$

Since $W_{k+1} \subset V_k$, this gives

$$\|f\|_{B_{k+1}} \leq \|f\|_{V_k}^{\beta} \|f\|_{\Omega}^{1-\beta} \leq \|f\|_{W_0}^{\beta\alpha_k} \|f\|_{\Omega}^{\beta(1-\alpha_k)} \|f\|_{\Omega}^{1-\beta}$$

$$= \|f\|_{W_0}^{\alpha_{k+1}} \|f\|_{\Omega}^{1-\alpha_{k+1}} ,$$

where $\alpha_{k+1} = \alpha_k \beta < \alpha_k$. Since $\|f\|_{W_0}^{\alpha} \|f\|_{\Omega}^{1-\alpha}$ increases as α decreases, we also have

$$\|f\|_{V_k} \leq \|f\|_{W_0}^{\alpha_{k+1}} \|f\|_{\Omega}^{1-\alpha_{k+1}}$$

and we obtain

$$\|f\|_{V_{k+1}} \leq \|f\|_{W_0}^{\alpha_{k+1}} \|f\|_{\Omega}^{1-\alpha_{k+1}} .$$

This proves (*) for all $k = 0, 1, \ldots, m$. The inequality with $k = m$ gives us the proposition since $V_m \supset \Omega_1$ and $W_0 \subset \Omega_0$.

Proposition 2 (Schwarz's lemma). Let $p_0 : \Omega \to \mathbb{C}^n$ be a connected domain, $\Omega_0 \subset\subset \Omega$ and $a \in \Omega_0$. Then, there is τ, $0 < \tau < 1$ such that if $f \in \mathcal{H}(\Omega)$ has a zero of order p at a (i.e., $D^{\alpha} f(a) = 0$ for $|\alpha| < p$), we have

$$\|f\|_{\Omega_0} \leq \tau^p \|f\|_{\Omega} \qquad \text{for any } f \in \mathcal{H}(\Omega).$$

Proof. We first prove the following

Let $B(R) = \{z \in \mathbb{C}^n \mid \|z\| < R\}$ and $0 < r < R$. If $f \in \mathcal{H}(B(R))$ has a zero of order p at 0, we have

$$\|f\|_{B(r)} \leq (\frac{r}{R})^p \|f\|_{B(R)} .$$

In fact, if $0 < |z| \leq r$, and $\varphi(u) = f(uz)$, then φ is holomorphic for $|u| < R/r = \rho$. Moreover $(d/du)^k \varphi(0) = 0$ for $k < p$. Hence $\frac{\varphi(u)}{u^p}$ is holomorphic for $|u| < \rho$. By the maximum principle, we have

$$|f(z)| = \frac{|\varphi(1)|}{|1^p|} \leq \lim_{\varepsilon \to +0} \sup_{|u|=\rho-\varepsilon} \left|\frac{\varphi(u)}{u^p}\right| \leq \left(\frac{r}{R}\right)^p \|f\|_{B(R)} .$$

To prove, now, the theorem, let $B_o \subset\subset B_1$ be balls with center a in Ω, and let $B_o \subset\subset \Omega_o$.

By what we have just proved, we have

$$\|f\|_{B_o} \leq \omega^p \|f\|_{\Omega} \quad \text{where } 0 < \omega < 1.$$

By Proposition 1, there is α, $0 < \alpha < 1$ so that

$$\|f\|_{\Omega_o} \leq \|f\|^{\alpha}_{B_o} \|f\|^{1-\alpha}_{\Omega} .$$

This gives

$$\|f\|_{\Omega_o} \leq \omega^{\alpha p} \|f\|^{\alpha}_{\Omega} \|f\|^{1-\alpha}_{\Omega} = \tau^p \|f\|_{\Omega} , \quad \tau = \omega^{\alpha} .$$

<u>Definition 1.</u> Let $P(z) = \sum_{|\alpha| \leq N} c_{\alpha} z^{\alpha}$ be a polynomial in n variables of degree N. We say that P is normalized if

$$\max_{\alpha} |c_{\alpha}| = 1 .$$

<u>Proposition 3.</u> Let K be a compact set in $\mathbb{C}^n$. Then there is a constant $C = C(K, n) > 0$ such that the following holds.

Let P be a normalized polynomial which is of degree $\leq d$ with respect to each variable $z_1, \ldots, z_n$. Let $0 < t < 1$ and let

$$S = S(t, P, K) = \{z \in K \mid |P(z)| \leq t^d\}.$$

Then

$$m(S) \leq Ct^{2/n} .$$

where m denotes Lebesgue measure in $\mathbb{C}^n$.

We shall need the following lemma.

Lemma 1. Let

$$q(t) = t^p + \sum_{k=1}^{p} a_k t^{p-k}, \quad a_j \in \mathbb{R}$$

be a monic polynomial of degree p with real coefficients. Then

$$\max_{-1 \le t \le +1} |q(t)| \ge 2^{-p}.$$

Proof. Let $g(\theta) = q(\cos\theta)$, $-\pi \le \theta \le \pi$. Now

$$g(\theta) = q\left(\frac{e^{i\theta} + e^{-i\theta}}{2}\right) = \sum_{k=-p}^{p} c_k e^{ik\theta}, \quad c_p = c_{-p} = 2^{-p}$$

Hence

$$2^{-p} = \left|\frac{1}{2\pi}\int_{-\pi}^{\pi} g(\theta)e^{-ip\theta}\,d\theta\right| \le \sup_{-\pi \le \theta \le \pi} |g(\theta)| = \sup_{-1 \le t \le 1} |q(t)|.$$

Proof of Proposition 3. Part 1. Case n = 1. Let $R > 0$ be so that $K \subset \{z \in \mathbb{C} \mid |z| \le R\}$. Let $\lambda_1, \ldots, \lambda_p$ be the zeros of P with $|\lambda_j| < R+1$, and $\mu_1, \ldots, \mu_q$ the zeros with $|\mu_j| \ge R+1$ ($p+q = d$, and we count zeros with multiplicity). Then

$$\begin{aligned} P(z) &= a(z-\lambda_1)\ldots(z-\lambda_p)\left(1-\frac{z}{\mu_1}\right)\ldots\left(1-\frac{z}{\mu_q}\right), \quad a \in \mathbb{C} \\ &= \sum_{\nu=0}^{d} c_\nu z^\nu, \quad \max|c_\nu| = 1. \end{aligned}$$

Let k be such that $|c_k| = 1$. Clearly, if A_k denotes the coefficient of z^k in

$$|a|(z+R+1)^d$$

we have (since $|\lambda_j| < R+1$, $|\mu_j| \ge R+1 > 1$), $1 = |c_k| \le |A_k| = \binom{d}{k}(R+1)^k|a|$. Since $\binom{d}{k} \le 2^d$, this gives

$$|a| \ge c^{-d}, \quad c = 2(R+1).$$

Since, for $|z| \le R$, we have $\left|1 - \frac{z}{\mu_1}\right| \ge \frac{1}{R+1}$, this implies that

$$S(t, P, K) \subset S_1 = \{z \in \mathbb{C} \mid |z| \le R, \ |z-\lambda_1|\ldots|z-\lambda_p| \le \tau^d\}$$

where $\tau = (R+1)\cdot c \cdot t = c_1 t$, say. Now, if $c_1 t \geq 1$, we have

$$(1) \qquad m(S_1) \leq \pi R^2 \leq c_1^2 \pi R^2 t^2 = c_2 t^2 , \quad c_2 = \pi c_1^2 R^2 .$$

Suppose that $\tau < 1$. Then $\tau^d \leq \tau^p$, so that

$$S_1 \subset S_0 = \{z \in \mathbb{C} \mid |z| \leq R, \ |z-\lambda_1| \ldots |z-\lambda_p| \leq \tau^p\}.$$

Let $\alpha_j = \operatorname{Re}\lambda_j$ and let $A = \{x \in \mathbb{R} \mid |x| \leq R, \ |x-\alpha_1| \ldots |x-\alpha_n| \leq \tau^p\}$. Then A contains the projection of S_0 onto the real axis $\mathbb{R}$. Let $m_0(A)$ be Lebesgue measure of A relative to $\mathbb{R}$. Clearly, A contains a nonempty open set, so that $m_0(A) > 0$. Consider the map $\varphi\colon \mathbb{R} \to \mathbb{R}$ defined by

$$\varphi(x) = -1 + \frac{2}{m_0(A)} m_0(\{y \in A \mid y \leq x\}).$$

Clearly $\varphi(x) \geq -1$ and $\varphi(-R) = -1$, $\varphi(+R) = 1$. Moreover, if $x < x'$

$$(*) \quad \varphi(x') - \varphi(x) = \frac{2}{m_0(A)} m_0(\{y \in A \mid x < y \leq x'\}) \leq \frac{2(x'-x)}{m_0(A)} ,$$

so that, in particular, φ is continuous. Moreover, if I is any open interval contained in $\mathbb{R} - A$, φ is clearly constant on I. Hence

$$\varphi(A) = \varphi(\mathbb{R}) = [-1,+1].$$

In view of $(*)$, we have, for $x \in A$,

$$|\varphi(x) - \varphi(\alpha_1)| \ldots |\varphi(x) - \varphi(\alpha_p)| \leq \left(\frac{2}{m_0(A)}\right)^p |x-\alpha_1| \ldots |x-\alpha_p| \leq \left(\frac{2\tau}{m_0(A)}\right)^p.$$

Since $\varphi(A) = [-1,+1]$, Lemma 1 shows that

$$2^{-p} \leq \sup_{x \in A} |\varphi(x) - \varphi(\alpha_1)| \ldots |\varphi(x) - \varphi(\alpha_p)| \leq (2\tau/m_0(A))^p ,$$

which gives $m_0(A) \leq 4\tau$.

In the same way, one shows that the projection A' of S_0 on the imaginary axis has measure

$$m_0(A') \leq 4\tau.$$

Hence

$$m(S_o) \leq 16\tau^2 .$$

This, and (1), imply that

$$m(S(t, P, K)) \leq c_3 t^2 , \quad c_3 = \max(c_2, 16c_1^2).$$

Part 2. The general case. We proceed by induction; suppose the result proved in $\mathbb{C}^{n-1}$.

Let $P(z) = \sum c_\alpha z^\alpha = \sum_{\beta \in \mathbb{N}^{n-1}} z'^\beta P_\beta(z_n)$, $z' = (z_1, \dots, z_{n-1})$.

There is then β_o such that P_{β_o} is a normalized polynomial.

Let

$$S_1 = \{z \in S(t, P, K) \mid |P_{\beta_o}(z_n)| \leq t^{d/n}\},$$

$$S_2 = S(t, P, K) - S_1 .$$

It follows at once from Part 1 above that

$$m(S_1) \leq M \cdot c_3 t^{2/n} = c_4 t^{2/n} ,$$

where M is the Lebesgue measure in $\mathbb{C}^{n-1}$ of the projection of K onto $\mathbb{C}^{n-1}$.

Let K_o be the projection of K onto the z_n-axis, K' that onto $\mathbb{C}^{n-1}$ and let $\zeta \in K_o$. Let $(z', \zeta) \in S_2$, and

$$Q(z') = P(z', \zeta)/P_{\beta_o}(\zeta).$$

Then $|P_{\beta_o}(\zeta)| \geq t^{d/n}$ and the maximum of the absolute values of the coefficients of Q is ≥ 1. Hence

$$S_{2,\zeta} = \{z' \in \mathbb{C}^{n-1} \mid (x, \zeta) \in S_2\} \subset \{z' \in K' \mid |Q(z')| \leq t^{d(n-1)/n}\} .$$

By induction hypothesis, there is $c_5 > 0$ so that the Lebesgue measure in $\mathbb{C}^{n-1}$ of $S_{2,\zeta}$ is $\leq c_5 t^{2/n}$. Hence, by Fubini's theorem

$$m(S_2) \leq c_6 t^{2/n} .$$

Hence

$$m(A) = m(S_1 \cup S_2) \leq (c_4 + c_6)t^{2/n} .$$

Remarks. (1). This inequality is essentially the best possible.

(2) For the actual application in view, a much weaker inequality, that can be proved using Jensen's inequality, is sufficient. We only need to know that the measure

$$m(S(t, P, K)) \to 0 \quad \text{as} \quad t \to 0$$

uniformly in P (in particular, uniformly with respect to $\deg P$). We have given the best possible inequality above in view of the interest of the methods used (which are due to Bishop).

Proposition 4. Let $p_o : \Omega \to \mathbb{C}^n$ be a connected domain over $\mathbb{C}^n$ and K a compact set in Ω. Let $f \in \mathcal{H}(\Omega)$. Then, there is θ, $0 < \theta < 1$ (depending only on K and f) such that if d, D are sufficiently large integers $D \leq d$, there is a polynomial $P(z_1, \ldots, z_n, w)$ of degree $\leq d$ in z_j, $j = 1, \ldots, n$, of degree $\leq D$ in w, such that, if $p_o(x) = (x_1, \ldots, x_n)$, $x \in \Omega$, we have

$$|P(x_1, \ldots, x_n, f(x))| \leq \theta^{d \cdot \delta} \quad , \quad \delta = D^{1/n} \, , \; x \in K.$$

Proof. Consider the vector space V of polynomials in $(z_1, \ldots, z_n, w)$, of degree $\leq d$ in z_j, $j = 1, \ldots, n$, of degree $\leq D$ in w. V is of dimension $(d+1)^n(D+1)$. If $(a_1, \ldots, a_n, b) \in \mathbb{C}^{n+1}$, and if N is an integer such that $N^n < (d+1)^n(D+1)$, then there is a normalized polynomial $P \in V$ such that $P(x, f(x))$ has a zero of order $\geq N$ at a [we write (z, w) for $(z_1, \ldots, z_n, w)$, etc].

It is sufficient to find a nonzero P with this property (we can then divide by the maximum absolute value of the coefficients). For this, we may take any nonzero element of the intersection of the kernels of the linear forms on V given by

$$P \mapsto \frac{\partial^{\alpha_1+\cdots+\alpha_n} P(x, f(x))}{\partial x_1^{\alpha_1} \cdots \partial x_n^{\alpha_n}}(a) \quad , \quad \alpha_1 + \cdots + \alpha_n \leq N-1.$$

[There are at most $N^n < \dim V$ of these forms.]

Let Ω_o be a connected neighborhood of K, $\Omega_o \subset\subset \Omega$, and let $M > 0$ be such that $|p_o(x)| < M$, $|f(x)| < M$ for $x \in \Omega_o$. If P is a normalized polynomial, $P \in V$, we have

$$|P(p_o(x), f(x))| \leq \sum_{\alpha_1, \ldots, \alpha_{n+1}} M^{\alpha_1 + \cdots + \alpha_{n+1}} = (1+M)^{nd+D} < C^d$$

(since $D \leq d$); here $0 \leq \alpha_1, \ldots, \alpha_n < d$, $0 \leq \alpha_{n+1} \leq D$. If now P is so chosen as to have a zero of order $N \geq (d+1)(D+1)^{1+n} - 1$ $(> dD^{1/n})$ at $(p_o(x_o), f(x_o))$ where $x_o \in K$ (as is possible by our remark above since we can choose an integer N with $(d+1)(D+1)^{1/n} - 1 \leq N < (d+1)(D+1)^{1/n} = \dim(V)^{1/n}$), we have, by Schwarz's lemma (Proposition 2),

$$|P(p_o(x), f(x))| \leq \tau^N C^d \qquad (x \in K)$$

where $0 < \tau < 1$ depends only on K and Ω_o. Since $C^{d/N} \to 1$ as $d, D \to \infty$ (since $N \geq dD^{1/n}$), we can choose θ, $0 < \theta < 1$, so that $\tau C^{d/N} < \theta$ for $d \geq D \geq D_o$. This gives

$$|P(x_1, \ldots, x_n, f(x))| < \theta^{d\delta} \quad , \quad \delta = D^{1/n} \; , \; x \in K.$$

Lemma 2. Suppose that $p_o : \Omega \to \mathbb{C}^n$ is a domain of holomorphy which is not $\mathcal{H}(\Omega)$-convex. Then, there is a compact set $K \subset \Omega$ and an infinite sequence $\{x_\nu\}$, $x_\nu \in \hat{K}$ such that the following holds:

There is $a \in \mathbb{C}^n$ and $\rho > 0$ such that $p_0(x_\nu) = a$ for all ν, $d(\{x_\nu\}) \geq \rho$ and the polydisc $P(x_\nu, \rho)$ of radius ρ about x_ν is contained in $\hat{K}$.

Proof. Let L be a compact set in Ω such that $\hat{L}$ is not compact. Let $\{y_\nu\}$ be a sequence of points in $\hat{L}$ without any limit point in Ω. Since $p_0 = (p_1, \dots, p_n)$ where the $p_j \in \mathcal{H}(\Omega)$, we have

$$|p_j(y_\nu)| \leq \|p_j\|_L .$$

Hence, by passing to a subsequence, if necessary, we may suppose that $p_0(y_\nu) \to z_0 \in \mathbb{C}^n$. Now, by chapter 7, Theorem 1, we have

$$d(\hat{L}) = d(L) = 2\eta > 0.$$

Let $\varepsilon > 0$ be small enough, and $a \in \mathbb{C}^n$, $|a-z_o| < \varepsilon$. Then the polydisc $P(y_\nu, \frac{1}{2}\eta)$ contains a point x_ν with $p_0(x_\nu) = a$, and we have $d(x_\nu) \geq 2\eta - \varepsilon$.

Let $K = L(\eta)$ be the union of the closures of the polydiscs of radius η about points of L. Then K is compact, and we have, by chapter 7, Proposition 2,

$$\hat{L}(\eta) \subset \hat{K}.$$

In particular, $P(y_\nu, \eta) \subset \hat{K}$, so that $P(x_\nu, \frac{1}{2}\eta) \subset \hat{K}$. We have only to set $\rho = \frac{1}{2}\eta$.

Theorem 1 (K. Oka). Any domain of holomorphy $p_0 : \Omega \to \mathbb{C}^n$ is $\mathcal{H}(\Omega)$-convex.

Proof. (E. Bishop). Suppose the result false. Then, we can choose K compact in Ω and $\{x_\nu\} \subset \hat{K}$ with the properties of Lemma 2. Let $f \in \mathcal{H}(\Omega)$, and d, D be large integers, $d \geq D$. Let $P_d(z_1, \dots, z_n, w)$ be a normalized polynomial of degree $\leq d$ in the z_j, of degree $\leq D$ in w such that

$$|P_d(x_1,\ldots,x_n,f(x))| < \theta^{dD^{1/n}}, \quad x \in K \quad (0 < \theta < 1,\ \theta = \theta(K,f)).$$

We write

$$P_d(z,w) = \sum_{k=0}^{D} P_k^{(d)}(z)w^k .$$

Let $X = \{(z,w) \in \mathbb{C}^n \mid (z,w) = (p_0(x), f(x)),\ x \in K\}$. Then

$$|P_d(z,w)| < \theta^{d\delta}, \quad (z,w) \in X, \quad \delta = D^{1/n} .$$

Consider the polydisc

$$Q = \{z \in \mathbb{C}^n \mid |z-a| < \rho\} \qquad (a, \rho \text{ as in Lemma 2}),$$

and let

$$S_{d,D} = \{z \in Q \mid \max_{k=0,\ldots,D} |P_k^{(d)}(z)| \leq \theta^{\frac{1}{2}d\delta}\} .$$

Since P_d is normalized, so is at least one of the $P_k^{(d)}$. Hence, by Proposition 3,

$$m(S_{d,D}) \leq c \cdot \theta^{\delta/n}, \quad c = c(Q).$$

In particular, if D is large enough, and

$$A_{d,D} = Q - S_{d,D} ,$$

we have

$$m(A_{d,D}) \geq \kappa > 0$$

where κ is independent of d, D, if $d \geq D$ and D is large enough.

We now keep D fixed, and set

$$A_p = \bigcup_{d \geq p} A_{d,D} , \quad A = \bigcap_{p=D}^{\infty} A_p .$$

Then

$$m(A) \geq \kappa .$$

Further, if $z \in A$, then, there are infinitely many d such that $z \notin S_{d,D}$. Hence, for $z \in A$, there are infinitely many d such that

$$\max_{k=0,\ldots,D} |P_k^{(d)}(z)| > \theta^{\frac{1}{2}d\delta} .$$

Hence, if we set $c_k^{(d)}(z) = P_k^{(d)}(z) / \max_{\ell=0,\ldots,D} |P_\ell^{(d)}(z)|$, $z \in A$, we have the following:

(*) For $(z, w) \in X$, $z \in A$, we have

$$\left|\sum_{k=0}^{D} c_k^{(d)}(z) w^k\right| < \theta^{\frac{1}{2}d\delta} \quad \text{for infinitely many } d.$$

Moreover, $\max_k |c_k^{(d)}(z)| = 1$. Hence, choosing a subsequence of the d for which (*) is true, we may suppose that $c_k^{(d)}(z) \to c_k$, $\max_k |c_k| = 1$. Hence, we obtain from (*) the following:

For $z \in A$, $(z, w) \in X$, there exist $c_k \in \mathbb{C}$, $k = 0, \ldots, D$, not all 0, $c_k = c_k(z)$, such that

$$\sum_{k=0}^{D} c_k \cdot w^k = 0.$$

From the definition of X we obtain the following result:

Lemma 3. There exists a subset $A \subset Q$ of positive measure such that if $z \in A$, then f takes at most D values on the set $p_o^{-1}(z) \cap \hat{K}$.

(For, by our remark above, any value of f on $p_o^{-1}(z) \cap \hat{K}$ satisfies an equation $\sum_{k=0}^{D} c_k w^k = 0$, where not all the c_k are 0.)

Let $\{x_\nu\}$ be an infinite sequence of points in $\hat{K}$ as in Lemma 2. Then $P(x_\nu, \rho) \subset \hat{K}$. As in Part 1 of the proof of Theorem 2 in chapter 7, we can find $f \in \mathcal{H}(\Omega)$ such that $f(x_\nu) \neq f(x_\mu)$ if $\nu \neq \mu$. Let now, for $z \in Q$, $y_\nu(z)$ be the point of $P(x_\nu, \rho)$ with $p_o(y_\nu(x)) = z$. We assert that the set of $z \in Q$ such that f does not separate the points $y_\nu(z)$ is of measure 0.

This would contradict Lemma 3 above, and therefore would complete the proof of Oka's theorem.

To prove the above assertion, we set

$$g_\nu = f \circ (p_o | P(x_\nu, \rho))^{-1}.$$

Then $g_\nu \in \mathcal{H}(Q)$. We set $A_{\mu,\nu} = \{z \in Q | \; g_\nu(z) = g_\mu(z)\}$. Clearly, $A_f = \bigcup_{\mu \neq \nu} A_{\mu,\nu}$. Moreover, since $g_\nu(a) = f(x_\nu) \neq f(x_\mu) = g_\mu(a)$, $(\mu \neq \nu)$, $A_{\mu,\nu}$ is an analytic set in Q, $\neq Q$. To prove that A_f is of measure 0, it suffices to show that each $A_{\mu,\nu}$ is of measure 0. This is an immediate consequence of the following lemma.

Lemma 4. Let Ω be an open connected set in $\mathbb{C}^n$ and $h \in \mathcal{H}(\Omega)$, $h \not\equiv 0$. Then, the set

$$Z = Z_h = \{z \in \Omega | \; h(z) = 0\}$$

has measure 0.

Proof of Lemma 4. It is sufficient to prove that any $a \in Z$ has a neighborhood U such that $U \cap Z$ is of measure 0. By a linear change of coordinates, we may suppose that $a = 0$ and that, for a small enough $r > 0$,

$$|h(0, \ldots, 0, z_n)| \neq 0 \quad \text{for} \quad |z_n| = r.$$

Then, for $\varepsilon > 0$ small enough, we have $h(z_1, \ldots, z_{n-1}, z_n) \neq 0$ for $|z_j| \leq \varepsilon$, $j = 1, \ldots, n-1$, $|z_n| = r$. Let $U = \{z \in \mathbb{C}^n | \; |z_j| < \varepsilon$, $j = 1, \ldots, n-1$, $|z_n| < r\}$. Then, for fixed $(z_1^o, \ldots, z_{n-1}^o)$, $|z_j^o| < \varepsilon$, the set

$$U \cap Z \cap \{z_j = z_j^o \, , \; j = 1, \ldots, n-1\}$$

is a finite set, hence of measure 0 in $\mathbb{C}$. It follows at once from Fubini's theorem that $U \cap Z$ is of measure 0 in $\mathbb{C}^n$.

This lemma proves our assertion, and, with it, Oka's theorem.

Oka's theorem is a special case of results proved in [22]. For one version of Oka's method, see [17].

The method of Bishop was given by him in [2].

9
AUTOMORPHISMS OF BOUNDED DOMAINS: CARTAN'S THEOREM

We have seen, in chapter 5, that if D is a bounded domain in $\mathbb{C}^n$, the group $G = \mathrm{Aut}(D)$ of analytic automorphisms of D is locally compact and acts properly on D. This chapter is devoted to proving a beautiful theorem due again to H. Cartan, that this group carries the structure of a Lie group and acts analytically on D.

The proof in this chapter is due to H. Cartan [7].

Let X be a Hausdorff space and $\mathfrak{U} = (U_i)_{i \in I}$ an open covering of X. Let $\varphi_i : U_i \to \Omega_i$ be a homeomorphism of U_i onto an open set $\Omega_i \subset \mathbb{R}^p$. Suppose that for any pair of elements $i, j \in I$, the map

$$\varphi_j \circ \varphi_i^{-1} : \varphi_i(U_i \cap U_j) \to \varphi_j(U_i \cap U_j)$$

is real analytic.

We call (X, U_i, φ_i) a real analytic manifold of dimension p. We normally omit the reference to the (U_i, φ_i).

Let X and Y be real analytic manifolds defined by data (U_i, φ_i), (V_j, ψ_j) respectively. A continuous map $f: X \to Y$ is called (real) analytic if the following holds: For any $a \in X$ and j such that $f(a) \in V_j$, there is a neighborhood U of a and an i such that $U \subset U_i$, $f(U) \subset V_j$ and such that the map $\psi_j \circ f \circ \varphi_i^{-1} : \varphi_i(U) \to \psi_j(V_j)$ is real analytic.

Clearly $\mathbb{R}^p$ has a canonical structure of real analytic manifold. Analytic mappings of a real analytic manifold X into $\mathbb{R}$ (or $\mathbb{C}$) are called real (or complex) valued real analytic functions on X.

Definition 1. Let G be a group which is also a real analytic manifold. G is called a Lie group if the mapping

$$G \times G \to G$$

given by $(x, y) \mapsto x \cdot y^{-1}$ is real analytic.

§1. Vector fields and Lie's theorem.

By a vector field on an open set $\Omega \subset \mathbb{R}^n$ we mean an n-tuple $X = (X_1, \ldots, X_n)$ of real-valued functions X_j on Ω. X is called C^∞ (respectively, real analytic) if the X_j are. If f is any C^∞ function on Ω we define

$$X(f)(x) = \sum_{j=1}^{n} X_j(x) \frac{\partial f}{\partial x_j}(x) , \quad x \in \Omega .$$

Moreover, the function $f \to X(f)$ on $C^\infty(\Omega)$ determines X completely.

If X, Y are two C^∞ vector fields, we define a third

$$Z = [X, Y]$$

by the relation

$$Z(f) = X(Y(f)) - Y(X(f)) , \quad f \in C^\infty(\Omega) .$$

It is easily verified that

$$Z = (Z_1, \ldots, Z_n) \text{ with } Z_j = \sum_{k=1}^{n} (X_k \frac{\partial Y_j}{\partial x_k} - Y_k \frac{\partial X_j}{\partial x_k}) .$$

We shall need the following result from the theory of ordinary differential equations. We shall not prove this here. A proof is given, e.g. in [21].

Proposition 1. Let Ω be an open set in $\mathbb{R}^p$ and $\Omega_o \subset\subset \Omega$ be open. Let X be a real analytic vector field on Ω. Then there exists

$\rho > 0$ and a unique real analytic map $g = g_X : \Omega_o \times I \to \Omega$, $I = \{ t \in \mathbb{R} \mid |t| < \rho \}$ such that

$$\frac{\partial g(x,t)}{\partial t} = X(g(x,t)), \; g(x,0) = x, \quad x \in \Omega_o, \; t \in I.$$

If U is open in $\mathbb{R}^q$ and $X : \Omega \times U \to \mathbb{R}^p$ is real analytic, then for $U_o \subset\subset U$, there is $\rho > 0$ such that the following holds:

For $\alpha \in U$, set $X_\alpha(x) = X(x, \alpha)$, $x \in \Omega$. Then there is an analytic map

$$g : \Omega_o \times I \times U_o \to \Omega \qquad (I = \{ t \in \mathbb{R} \mid |t| < \rho \})$$

such that the map $g_\alpha : \Omega_o \times I \to \Omega$ defined by

$$g_\alpha(x,t) = g(x,t,\alpha)$$

satisfies $g_\alpha = g_{X_\alpha}$. Moreover, if $t, s, t+s \in I$, we have

$$g_\alpha(g_\alpha(x,t), s) = g_\alpha(x, t+s)$$

whenever $\alpha \in U_o$, $x \in \Omega_o$ and $g_\alpha(x,t) \in \Omega_o$.

We call $g = g_X$ the local one-parameter group associated to the vector field X.

Note that for any $f \in C^\infty(\Omega)$, we have

$$X(f)(x) = \lim_{t \to 0} t^{-1} \{ f \circ g_X(x,t) - f(x) \} .$$

Definition 2. Let V be a finite dimensional vector space of vector fields on an open set $\Omega \subset \mathbb{R}^p$. We say that V is a Lie algebra of vector fields if whenever $X, Y \in V$, we have $[X, Y] \in V$.

In all that follows, we shall assume that the vector fields belonging to V are real analytic.

We shall use the following result from classical Lie theory. Since it appears somewhat difficult to give a reference to the theorem in the form we need, we shall give a proof.

Theorem 1 (Lie's theorem). Let V be a finite-dimensional Lie algebra of real analytic vector fields on an open connected set $\Omega \subset \mathbb{R}^p$. Let $\Omega_0 \subset\subset \Omega$. Then there exists a neighborhood U of 0 in V and a real analytic map

$$g: \Omega_0 \times U \to \Omega$$

with the following properties.

Let $g_u: \Omega_0 \to \Omega$ be the map $x \mapsto g(x, u)$.

(i) For u and v sufficiently near 0, there is a unique $w = w(u, v) \in U$ such that

$$g_u \circ g_v = g_w \quad \text{on } \Omega_0 \cap g_v^{-1}(\Omega_0).$$

(ii) For $u \in U$, the map $t \mapsto g_{tu}$ ($t \in \mathbb{R}$, near 0) is the one-parameter group associated to the vector field u. (In particular, g_0 = identity.)

(iii) For u_0, v_0 sufficiently near 0, the maps

$$u \mapsto w(u, v_0) \quad \text{and} \quad v \mapsto w(u_0, v)$$

are analytic isomorphisms of a neighborhood of $0 \in V$ onto neighborhoods of v_0, u_0 respectively.

The map g is called the local Lie group of transformations associated to V.

Proof. Let $\Omega_0 \subset\subset \Omega_1 \subset\subset \Omega$ and let $X^1, \ldots, X^m$ be a basis of V. If $a \in \mathbb{R}^m$, we shall denote by $X^{(a)}$ the vector field $X^{(a)} = \sum_{j=1}^{m} a_j X^j$. Let $\rho > 0$ and $I = \{|t| < \rho\}$ be such that there is a map

$$\varphi: \Omega_1 \times I \times U_1 \to \Omega, \quad U_1 = \{a \in \mathbb{R}^m \mid |a_j| < \rho\}$$

such that

$$\frac{\partial \varphi}{\partial t}(x, t, a) = X^{(a)}(\varphi(x, t, a)), \quad \varphi(x, 0, a) = x. \quad \text{(Proposition 1).}$$

We suppose ρ so small that $\varphi(x, t, a) \in \Omega_1$ for $x \in \Omega_0$, $t \in I$, $a \in U_1$.

Let $X \in V$, $X = X^{(a)}$ and let $\sigma_{t,X}: \Omega_o \to \Omega$ be the map

$$\sigma_{t,X}(x) = \varphi(x,t,a).$$

For any $Y \in V$, we define a vector field $X_{t,*}(Y)$ on Ω_o as follows: Let $f \in C^\infty(\Omega)$. Then

$$X_{t,*}(Y)(f) = Y(f \circ \sigma_{-t,X})(\sigma_{t,X}(x)) .$$

[Note that if $Y = (Y_1, \ldots, Y_p)$ and $\sigma_{t,X} = (\sigma_{t,1}, \ldots, \sigma_{t,p})$, we have

$$X_{t,*}(Y) = (Z_1, \ldots, Z_p)$$

where

$$Z_j(x) = \sum_{k=1}^{p} \frac{\partial \sigma_{-t,j}}{\partial x_k}(y) Y_k(y) , \quad y = \sigma_{t,X}(x) . \;]$$

Let $Z(t) = X_{t,*}(Y)$, where $X, Y \in V$. We assert:

<u>Lemma a.</u> $\dfrac{dZ(t)}{dt} = X_{t,*}([X,Y])$.

<u>Proof of Lemma a.</u> Let $f \in C^\infty(\Omega)$. Then,

$$\frac{dZ}{dt}(t_o)(f) = \lim_{t \to 0} t^{-1}\{Y(h \circ \sigma_{-t,X}) \circ \sigma_{t,X} - Y(h)\} \circ \sigma_{t_o,X} \quad , \; h = f \circ \sigma_{-t_o,X}$$

$$= \lim_{t \to 0} \{t^{-1} Y(h \circ \sigma_{-t,X} - h) - t^{-1}[Y(h) \circ \sigma_{-t,X} - Y(h)]\} \circ \sigma_{t_o,X} .$$

Now, since $\sigma_{t,X}$ is the one parameter group associated to X, we have

$$\lim_{t \to 0} t^{-1}\{Y(h) \circ \sigma_{-t,X} - Y(h)\} = -X(Y(h)) .$$

Also, the function $\psi = h \circ \sigma_{-t,X} \in C^\infty(I \times \Omega_o)$. Hence, so is the function

$$F = \begin{cases} t^{-1}(h \circ \sigma_{-t,X} - h) = t^{-1}(\psi(t,x) - \psi(0,x)) & t \neq 0 \\ \dfrac{\partial \psi}{\partial t}(0,x) & t = 0 \end{cases} .$$

Hence

$$\lim_{t \to 0} t^{-1}\{Y(h \circ \sigma_{-t,X} - h) = Y(\lim_{t \to 0} t^{-1}(h \circ \sigma_{-t,X} - h)) = -YX(h).$$

Hence

$$\frac{dZ}{dt}(t_o)(f) = [X,Y](h)\circ\sigma_{t_o,X} = [X,Y](f\circ\sigma_{-t_o,X})\circ\sigma_{t_o,X}$$

$$= X_{t_o,*}([X,Y])(f) .$$

Lemma b. For $X, Y \in V$ and small t, $X_{t,*}(Y)$ is the restriction to Ω_o of an element of V.

Proof of Lemma b. Let $\psi(t) = X_{t,*}(Y)$, $x \in \Omega_o$. Clearly $\psi(0) = Y$. Now, by Lemma a, we have

$$\frac{d^k\psi}{dt^k}(0) = Y^{(k)}$$

where

$$Y^{(0)} = Y\ ,\ Y^{(k)} = [X, Y^{(k-1)}]\ ,\quad k \geq 1.$$

Now, $Y^{(k)}$ is the image of Y under the k-th iterate of the endomorphism $A: Z \to [X, Z]$ of V. Since ψ is real analytic, we have, for small enough t,

$$\psi(t) = \sum_{k=0}^{\infty} \frac{t^k}{k!}\,\frac{d^k\psi(0)}{dt^k} = e^{tA}(Y) .$$

Lemma c. If $X, Y \in V$ and t, s are small, we have

$$\sigma_{s,Y}\circ\sigma_{t,X} = \sigma_{t,X}\circ\sigma_{s,Y'}$$

where $Y' = X_{t,*}(Y)$.

Proof of Lemma c. Let $\Phi(s,t,x) = \sigma_{-t,X}\circ\sigma_{s,Y}\circ\sigma_{t,X}$. We verify at once, by differentiation that

$$\frac{\partial\Phi}{\partial s}(0,t,x) = X_{t,*}(Y)(x) = Y'(x).$$

Hence

$$\frac{\partial\Phi}{\partial s}(s,t,x) = \frac{\partial\Phi}{\partial s'}(0,t,\Phi(s,t,x)) = Y'(\Phi(s,t,x))$$

Moreover $\Phi(0,t,x) = x$. Hence $s \mapsto \Phi(s,t,x)$ is the one-parameter group associated to the vector field Y', i.e.,

$$\Phi(s,t,x) = \sigma_{s,Y'}(x).$$

Lemma d. If $X^1, \ldots, X^m$ is a basis for V and $a \in \mathbb{R}^m$ is small, define, for $x \in \Omega_o$

$$F(a,x) = g_{a_1 X^1} \circ \cdots \circ g_{a_m X^m}(x) ,$$

and let $F = (F_1, \ldots, F_p)$. Then, there is a map $u: U \to GL(m, \mathbb{R})$ of a neighborhood U of 0 in $\mathbb{R}^m$ such that

$$u(0) = I_m \quad \text{(the } m \times m \text{ identity matrix)}$$

and

$$\frac{\partial F}{\partial a_i}(a,x) = \sum_{j=1}^{m} u_{ij}(a) X^j(F(a,x)) , \qquad u = (u_{ij}).$$

Proof of Lemma d. If $h \in \mathbb{R}$ is small, and $a' = (a_1, \ldots, a_i + h, \ldots, a_m)$, $1 \le i \le m$, we have, from Lemma c,

$$F(a',x) - F(a,x) = \sigma_{h, Z_i}(y) - y, \qquad y = F(a,x),$$

where

$$Z_i = Z_i^{(a)} = X^1_{-a_1, *} \cdots X^{i-1}_{-a_{i-1}, *}(X^i).$$

Hence

$$\frac{\partial F}{\partial a_i}(a,x) = Z_i(y) = Z_i(F(a,x)) .$$

Furthermore, the map $a \to Z_i^{(a)}$ is analytic. Hence, there are unique analytic functions u_{ij} such that

$$Z_i^{(a)} = \sum_{j=1}^{m} u_{ij}(a) X^j ,$$

so that

$$\frac{\partial F}{\partial a_i}(a,x) = \sum_{j=1}^{m} u_{ij}(a) X^j(F(a,x)) .$$

Moreover, $Z_i^{(0)} = X^i$, so that $u_{ij}(0) = \delta_{ij}$ ($= 1$ if $i = j$, 0 if $i \neq j$). This proves Lemma d.

Let $u: U \to GL(m, \mathbb{R})$ be the matrix constructed in Lemma d and let $v: U \to GL(m, \mathbb{R})$ be the inverse: $v(a) = u(a)^{-1}$. Let v^j be the vector field

$$a \mapsto (v_{j1}(a), \ldots, v_{jm}(a)) .$$

For $\alpha = (\alpha_1, \ldots, \alpha_m)$ sufficiently near 0 and $U_o \subset\subset U$, consider the local one-parameter group

$$\gamma_{t,\alpha}: \frac{\partial \gamma_{t,\alpha}}{\partial t}(a) = \sum_{j=1}^{m} \alpha_j v^j(\gamma_{t,\alpha}(a)) , \quad \gamma_{0,\alpha}(a) = a \quad \text{for all } a \in U_o .$$

Then, there is a C^∞ function $\Gamma : W \times U_o \to U$ (W a suitable neighborhood of 0 in $\mathbb{R}^n$) so that

$$\Gamma(t\alpha, a) = \gamma_{t,\alpha}(a) .$$

In fact, if $t\alpha = s\beta$, both the functions $\tau \mapsto \gamma_{\tau t,\alpha}$ and $\tau \mapsto \gamma_{\tau s,\beta}$ satisfy the equation

$$\frac{\partial \gamma}{\partial \tau} = \sum_{j=1}^{m} (t\alpha_j) v^j(\gamma) , \quad \gamma(0) = a$$

and so are identical.

Consider now the map

$$f : \alpha \mapsto \Gamma(\alpha, 0) .$$

We have, for all small α,

$$\sum_{j=1}^{m} \alpha_j \frac{\partial f}{\partial \alpha_j}(0) = \frac{\partial}{\partial t} \Gamma(t\alpha, 0)\Big|_{t=0} = \sum_{j=1}^{m} \alpha_j v^j(0) ,$$

so that

$$\frac{\partial f}{\partial \alpha_j}(0) = v^j(0) .$$

It follows that there are neighborhoods $W_o \subset\subset W$ and $U_2 \subset\subset U_o$ of 0 in $\mathbb{R}^m$ such that f is an analytic isomorphism of W_o onto U_2. As above, there is an analytic map

$G : U_3 \times \Omega_o \to \Omega$, $U_3 \subset\subset U_2$ being a suitable neighborhood of 0 such that

$$\frac{\partial G(ta,x)}{\partial t} = \sum_{j=1}^{m} a_j X^j(G(ta,x)) , \quad G(0,x) = x, \quad x \in \Omega_o .$$

If $F(a,x) = g_{a_1, X^1} \circ \dots \circ g_{a_m, X^m}(x)$ as above, we claim that for α, b sufficiently close to 0 in $\mathbb{R}^m$, we have

$$(*) \qquad G(\alpha, F(b,x)) = F(c,x), \qquad c = \Gamma(\alpha, b) .$$

In fact, if $G(t) = G(t\alpha, F(b,x))$, $F(t) = F(\Gamma(t\alpha, b), x)$, we have

$$\text{(i)} \qquad G(0) = F(b,x) , \quad \frac{dG}{dt} = \sum_{j=1}^{m} \alpha_j X^j(G).$$

On the other hand

$$\text{(ii)} \qquad F(0) = F(\Gamma(0,b),x) = F(b,x) \quad \text{and}$$

$$\frac{dF}{dt} = \sum_{j=1}^{m} \frac{\partial F}{\partial a_j}(\Gamma) \frac{d\Gamma_j(t\alpha, b)}{dt} , \quad \Gamma = (\Gamma_1, \dots, \Gamma_m)$$

$$= \sum_{j=1}^{m} \frac{\partial F}{\partial a_j}(\Gamma) \sum_{k=1}^{m} \alpha_k v_{kj}(\Gamma(t\alpha, b))$$

$$= \sum_{i,j,k=1}^{m} \alpha_k v_{kj}(\Gamma(t\alpha, b)) u_{ji}(\Gamma(t\alpha, b)) X^i(F(t)) \qquad \text{[Lemma d]}$$

$$= \sum_{k=1}^{m} \alpha_k X^k(F(t)) \qquad \text{since } u(a) = v(a)^{-1} .$$

The equation (*) follows from the uniqueness of solutions of ordinary differential equations.

If we set $b = 0$ in (*), we obtain

$$G(\alpha, x) = F(a,x) \quad \text{with } a = \Gamma(\alpha, 0) = f(\alpha).$$

Hence, if α, β are close enough to 0, we have

$$\begin{aligned} G(\alpha, G(\beta, x)) &= G(\alpha, F(b,x)) , & b &= f(\beta) , \\ &= F(c,x) , & c &= \Gamma(\alpha, b) = \Gamma(\alpha, f(\beta)) \\ &= G(\gamma, x) , & \gamma &= \gamma(\alpha, \beta) = f^{-1}(\Gamma(\alpha, f(\beta))). \end{aligned}$$

[Note that if α, β are close to 0, $\gamma(\alpha, \beta) \in U_2$.] Also, since

$\frac{\partial\Gamma(0,0)}{\partial\alpha_1}, \ldots, \frac{\partial\Gamma(0,0)}{\partial\alpha_m}$ are $\mathbb{R}$-independent, so are

$\frac{\partial\Gamma}{\partial\alpha_1}(\alpha, \beta), \ldots, \frac{\partial F}{\partial\alpha_m}(\alpha, \beta)$ for sufficiently small α, β. Hence the map $\alpha \mapsto \gamma(\alpha, \beta)$ is, for fixed β, an isomorphism of a neighborhood of 0 onto one of $\gamma(0, \beta) = \beta$. Again, $\Gamma(0, \beta) = \beta$, so that $\frac{\partial\Gamma}{\partial\beta_1}, \ldots, \frac{\partial\Gamma}{\partial\beta_m}$ are $\mathbb{R}$-independent for sufficiently small α, β, so that the map $\beta \to \gamma(\alpha, \beta)$ is, for fixed β, an isomorphism of a neighborhood of 0 onto one of $\gamma(\alpha, 0) = \alpha$. This proves all the assertions of Theorem 1 except the uniqueness assertion in (i). This latter is a consequence of the following.

Lemma e. Let $X^1, \ldots, X^m$ be linearly independent real analytic vector fields on an open connected set $\Omega \subset \mathbb{R}^p$. Let $\Omega_o \subset\subset \Omega$ and $I = \{t \in \mathbb{R} \mid |t| < \rho\}$, where ρ is sufficiently small. Then, there is a unique analytic map $F: \Omega_o \times U \to \Omega$ (U a suitable neighborhood of 0 in $\mathbb{R}^m$) so that

$$(**) \quad \frac{\partial F(x, ta)}{\partial t} = \sum_{j=1}^{m} a_j X^j(F(x, ta)), \; a = (a_1, \ldots, a_m), \; F(x, 0) = x.$$

Further, if U is small enough, the map

$$a \mapsto F_a, \; F_a(x) = F(x, a), \quad x \in \Omega_o$$

is injective.

Proof of Lemma e. Let $f: \Omega_o \times U \times I \to \Omega$ be an analytic map such that

$$\frac{\partial f}{\partial t}(x, a, t) = \sum_{j=1}^{m} a_j X^j(f), \quad f(x, a, 0) = x.$$

Then, if $ta = sb$, we have $f(x, a, t) = f(x, b, s)$ (both maps $\tau \mapsto f(x, a, t\tau)$, $\tau \mapsto f(x, b, s\tau)$ are solutions of the differential equation

$\frac{\partial f}{\partial t} = \sum_{j=1}^{m} ta_j X^j(f)$, $f(0) = x)$, so that there is an analytic map

$$F:\Omega_o \times U \to \Omega$$

satisfying (**). Also

$$\left.\frac{\partial F(ta,x)}{\partial t}\right|_{t=0} = \sum_{j=1}^{m} a_j \frac{\partial F}{\partial a_i}(0,x) = \sum_{j=1}^{m} a_j X^j(x) \text{ for all } a \in U,$$

so that

$$\frac{\partial F}{\partial a_j}(0,x) = X^j(x) .$$

We now assert that there exist finitely many points $x_1, \ldots, x_N \in \Omega_o$ such that the map $\Phi : U \to \mathbb{R}^{mN}$

$$(a_1, \ldots, a_m) \to (F(a,x_1), \ldots, F(a,x_N))$$

is such that the differential map

$$d\Phi_o : \mathbb{R}^m \to \mathbb{R}^{mN}$$

is injective. Since $\frac{\partial F}{\partial a_j}(0,x) = X^j(x)$, it is enough to find $x_1, \ldots, x_N \in \Omega_o$ such that

$$\sum_{j=1}^{m} a_j X^j(x_p) = 0 \;, \quad p = 1, \ldots, N \Rightarrow a_j = 0, \; j = 1, \ldots, m.$$

To do this, let V be the $\mathbb{R}$-vector space generated by $X^1, \ldots, X^m$ and let $x_1 \in \Omega_o$ be such that $X^1(x_1) \neq 0$ $(X^1 \not\equiv 0$ on Ω, hence $\not\equiv 0$ on $\Omega_o)$. Let $V_1 = \{X = \sum_{j=1}^{m} a_j X^j \mid X(x_1) = 0\}$. Then $\dim V_1 < \dim V = m$. If $V_1 = \{0\}$, the theorem is proved. If not, let $x_2 \in \Omega_o$ be so that $Y(x_2) \neq 0$ for some $Y \in V_1$. Let $V_2 = \{X \in V_1 \mid X(x_2) = 0\}$. Then $\dim V_2 < \dim V_1 \leq m-1$. Proceeding in this way, we find $x_1, \ldots, x_N \in \Omega_o$, $N \leq m$, such that

$$\{X \in V \mid X(x_1) = \ldots = X(x_N) = 0\}$$

has dimension 0. Then, if $\sum_{j=1}^{m} a_j X^j(x_p) = 0$, $p = 1, \ldots, N$, then $a_1 = \ldots = a_m = 0$.

It follows from the implicit function theorem that Φ is injective on a sufficiently small neighborhood of 0. This proves Lemma e.

§2. Cartan's theorem.

Let D be a bounded domain in $\mathbb{C}^n$ and G = Aut(D) the group of holomorphic automorphisms of D. When we speak of vector fields on D, we shall identify $\mathbb{C}^n$ with $\mathbb{R}^{2n}$. Let $X: D \to \mathbb{C}^n$ be a holomophic map. This can be interpreted as a vector field on D; if $X = (X_1, \ldots, X_n)$, the effect of X on a C^∞ function φ is given by

$$X(\varphi)(x) = \sum_{j=1}^{n} X_j(x) \frac{\partial \varphi}{\partial z_j}(x).$$

We say that the vector field X is associated to G if the following holds: Let $t \mapsto g_t$ be the local one-parameter group associated to X on $\Omega_o \subset\subset D$. Then for sufficiently small t, g_t is the restriction to Ω_o of an element of G.

It follows from the principle of analytic continuation that the choice of Ω_o is irrelevant.

Proposition 2. Let $GL(n, \mathbb{C})$ denote the group of invertible $n \times n$ complex matrices. Then G is homeomorphic to a closed subset of $D \times GL(n, \mathbb{C})$.

Proof. Let $x_o \in D$ and let $K = \{\sigma \in G \mid \sigma(x_o) = x_o\}$. By chapter 5, Proposition 6, K is compact. For $\sigma \in G$, let $J(\sigma)$ be the matrix $(\frac{\partial \sigma_i}{\partial z_j}(x_o))$, $\sigma = (\sigma_1, \ldots, \sigma_n)$. Clearly $J(\sigma) \in GL(n, \mathbb{C})$.

Consider the map

$$\varphi: G \to D \times GL(n, \mathbb{C})$$

given by

$$\varphi(\sigma) = (\sigma(x_o), J(\sigma)) .$$

We claim that φ is injective and proper. In fact, the map $\sigma \to \sigma(x_o)$ is a proper map of G into D (because of Proposition 6 of chapter 5). Hence so is φ. To prove that φ is injective, suppose that $\varphi(\sigma) = \varphi(\tau)$. Then $\tau^{-1} \circ \sigma \in K$. Also

$$J(\tau^{-1} \circ \sigma) = \left(\frac{\partial \tau_i^{-1}}{\partial z_j}(\sigma(x_o))\right)\left(\frac{\partial \sigma_i}{\partial z_j}(x_o)\right)$$

$$= J(\tau)^{-1} \cdot J(\sigma) \quad (\text{since } \sigma(x_o) = \tau(x_o))$$

$$= \text{identity} \quad \text{since } J(\sigma) = J(\tau) \text{ by hypothesis.}$$

Hence by chapter 5, Proposition 1, $\tau^{-1} \circ \sigma$ = identity, so that φ is injective.

Since φ is injective and proper, it is a homeomorphism onto a closed subset of $D \times GL(n, \mathbb{C})$.

Proposition 3. Let U_o be a non-empty open subset of D. Then for any $\Omega \subset\subset D$ and $\varepsilon > 0$, there is a $\delta > 0$ such that the following holds:

If $\sigma, \tau \in G$ and $|\sigma(x) - \tau(x)| < \delta$ for $x \in U_o$, we have $|\sigma(x) - \tau(x)| < \varepsilon$ for $x \in \Omega$.

Proof. This follows at once from Vitali's theorem, (chapter 1, Proposition 7).

Let now $X^1, \ldots, X^m$ be holomorphic maps of D into $\mathbb{C}^n$ and let V be the $\mathbb{C}$-vector space spanned by the X^j. Suppose that V is a Lie algebra of vector fields. [Note that with the standard identification of $\mathbb{C}$ with $\mathbb{R}^2$, this means simply that

$$[X^i, X^j] = (Z_1, \ldots, Z_n)$$

is a complex linear combination of the X^k; here

$$Z_s = \sum_{r=1}^{n} (X_r^i \frac{\partial X_s^j}{\partial z_r} - X_r^j \frac{\partial X_s^i}{\partial z_r}) \quad , \quad s = 1, \ldots, n .]$$

Let $\Omega_o \subset\subset D$ and $g: \Omega_o \times U \to D$ be the local Lie group of V, U being a suitable neighborhood of 0 in V. Suppose that for any $u \in U$, the map

$$g_u: \Omega_o \to D \ : \ g_u(x) = g(x, u)$$

is the restriction to Ω_o of an element $f_u \in G$. We define a map

$$f: D \times U \to D$$

by

$$f(x, u) = f_u(x).$$

Proposition 4. The map f is real analytic.

Proof. Let $U_o \subset\subset U$, and $\Omega_1 \subset\subset \Omega_o \subset\subset D$. Then $\bigcup_{u \in U_o} g_u(\Omega_1) = g(\Omega_1 \times U_o) \subset\subset D$. Let Ω be a connected open set with

$$g(\Omega_1 \times U_o) \subset\subset \Omega \subset\subset D.$$

Let W be a sufficiently small neighborhood of 0 in V and

$$h: \Omega \times W \to D$$

the local Lie group on Ω defined by V. For $u \in W$, $v \in U_o$, we have, with the notation as in Theorem 1

$$g_v \bullet h_u = g_{w(v,u)} \quad \text{on } \Omega_1$$

(since $h|\Omega_o \times W = g|\Omega_o \times W$). Hence, by the principle of analytic continuation, we have

$$f_v \circ h_u = f_{w(v,u)} \quad \text{on } \Omega.$$

Now, the map $u \to w(v, u)$ is an analytic isomorphism of a neighborhood N_o of 0 onto a neighborhood N of v. Let φ be the inverse. We have, for $x \in \Omega$, $w \in N$,

$$f_w(x) = f_v \circ h_{\varphi(w)}(x).$$

Hence

$$f(x, w) = f_v(h(x, \varphi(w))), \quad w \in N, \ x \in \Omega .$$

Since h is analytic on $\Omega \times W$ and f_v on D, it follows that f is analytic on $\Omega \times N$. Since U_o is <u>any</u> relatively compact subset of U and Ω <u>any</u> relatively compact connected set with $g(\Omega_1 \times U_o) \subset\subset \Omega$, it follows that f is analytic on $D \times U$.

<u>Corollary.</u> Let X be a vector field associated to G. Then there is an analytic map

$$g: D \times \mathbb{R} \to D$$

such that

$$\frac{\partial g(x,t)}{\partial t} = X(g(x,t)) \ , \quad g(x, 0) = x$$

and such that, for any $t \in \mathbb{R}$, the map $x \mapsto g(x,t)$ belongs to G.

<u>Proof.</u> Let $\Omega_o \subset\subset D$ and $\rho > 0$ be small enough. Let $I = \{t \in \mathbb{R} \mid |t| < \rho\}$. Then there is an analytic map

$$h: \Omega_o \times I \to D$$

such that

$$\frac{\partial h(x,t)}{\partial t} = X(h(x,t)) \ , \quad h(x,0) = x \ , \ x \in \Omega_o, \ t \in I,$$

and the map $h_t: x \to h(x,t)$ is the restriction to Ω_o of an element $f_t \in G$. By Proposition 4, the map

$$f: D \times I \to D$$

defined by $f(x,t) = f_t(x)$ is analytic. We define g as follows: Let $t_o \in \mathbb{R}$ and $p > 0$ be an integer such that $s_o = t_o/p \in I$. We set

$$g(x, t_o) = \underbrace{f_{s_o} \cdots f_{s_o}}_{p \text{ times}}(x).$$

That g is independent of the choice of p follows from the properties of the local one-parameter group stated in Proposition 1. That g satisfies the required differential equation follows from the principle of analytic continuation.

This map g is called the one-parameter group associated to X.

Theorem 2. Let $\{\sigma_\nu\}$ be a sequence of elements of $G = \mathrm{Aut}(D)$, D being a bounded domain in $\mathbb{C}^n$. Suppose that $\{\sigma_\nu\}$ converges to the identity element in G and that there exists a sequence $\{m_\nu\}$ of integers, $m_\nu \to \infty$ as $\nu \to \infty$ with the following property:

The sequence $\{X_\nu\}$, $X_\nu(x) = m_\nu(\sigma_\nu(x) - x)$, of maps of D into $\mathbb{C}^n$ converges, uniformly on compact subsets of D, to a map $X: D \to \mathbb{C}^n$.

Then X is a vector field associated to G.

Proof. Let $\Omega_0 \subset\subset D$ and let $\rho > 0$ be so small that the local one-parameter group

$$g: \Omega_0 \times I \to D$$

corresponding to X is defined on $\Omega_0 \times I$. Let $0 < t_0 < \rho$, and let q_ν be the largest integer $\leq m_\nu t_0$. Then, $q_\nu \to \infty$ and $0 \leq m_\nu t_0 - q_\nu < 1$. We shall prove that there exists a nonempty open set $B \subset \Omega_0$ such that if t_0 is small enough, then $\sigma_\nu^{q_\nu}$ converges uniformly on B to the map $g_{t_0}: x \mapsto g(x, t_0)$. It follows from Vitali's theorem (chapter 1, Proposition 7) and chapter 5, Theorem 4 that $\sigma_\nu^{q_\nu}$ converges to an element $\sigma \in G$ which, by the principle of analytic continuation satisfies $\sigma|\Omega_0 = g_{t_0}$.

We have

$$\left.\frac{\partial g(x,t)}{\partial t}\right|_{t=0} = X(g(x,0)) = X(x) \quad , \quad x \in \Omega_0 .$$

If K is a compact subset of Ω_0, with nonempty interior, this implies that

$$(1) \qquad g(x,t) - x = g(x,t) - g(x,0) = t\{X(x) + \varepsilon_1(x,t)\}$$

where $\varepsilon_1(x,t) \to 0$ as $t \to 0$, uniformly for $x \in K$. Moreover,

$$\sigma_\nu(x) - x = \frac{1}{m_\nu} X_\nu(x) = \frac{t_o}{q_\nu} X'_\nu(x), \quad X'_\nu = \frac{q_\nu}{t_o m_\nu} X_\nu \quad .$$

Now $X'_\nu \to X$ as $\nu \to \infty$ (since $0 \le m_\nu t_o - q_\nu < 1$). Hence

(1') $$\sigma_\nu(x) - x = \frac{t_o}{q_\nu} \{X(x) + \varepsilon_\nu(x)\}$$

where $\varepsilon_\nu(x) \to 0$ as $\nu \to \infty$, uniformly for $x \in K$. It follows at once that

(2) $$|g(x, t_o/q_\nu) - \sigma_\nu(x)| < \delta_\nu/q_\nu \quad , \quad x \in K ,$$

where $\delta_\nu \to 0$ as $\nu \to \infty$.

Let $M > 0$ be so chosen that $\|X\|_{\Omega_o} < M$ and let B be a ball of radius r about $x_o \in K$, $B \subset K$. We shall suppose that the ball of radius r about any point of B is contained in K. We assume that $t_o < r/M$.

We assert the following:

If ν is sufficiently large and $x \in B$, then

$$g(x, pt_o/q_\nu) \in K \quad , \quad \sigma_\nu^p(x) \in K$$

for $p = 1, \ldots, q_\nu$.

To prove this, let f denote either of the two mappings $x \mapsto \sigma_\nu(x)$, $x \mapsto g(x, t_o/q_\nu)$. Then, because of (1) and (1'), if ν is large enough, we have, for $x \in K$,

(3) $$|f(x) - x| < t_o M/q_\nu \; < \; r/q_\nu \quad .$$

Hence, for $x \in B$, $f(x) \in K \subset \Omega_o$. Thus $f_2(x) = f(f(x))$ is defined and, by (3) at the point $f(x)$, we have

$$|f_2(x) - f(x)| < r/q_\nu \; ,$$

so that

$$|f_2(x) - x| < 2r/q_\nu \le r \qquad (\text{if } q_\nu \ge 2) \; .$$

Thus, if $q_\nu \geq 2$, $f_2(x) \in K$ and we can continue the above iteration of the inequality (3). If we set $f_1(x) = f(x)$, $f_p(x) = f(f_{p-1}(x))$ whenever $f_{p-1}(x) \in K$, this gives us the following:

For $1 \leq p \leq q_\nu$, and $x \in B$, $f_p(x)$ is defined and

$$|f_p(x) - x| < pr/q_\nu \leq r.$$

In particular, $f_p(x) \in K$.

Let $x \in B$, $x_p = \sigma^p(x)$ and $y_p = g(x, pt_o/q_\nu)$, $p = 1, \ldots, q_\nu$. From (1), it follows that, for $y, y' \in K$ and t sufficiently small, we have

$$|g(y,t) - g(y',t)| \leq |y-y'|(1 + M|t|).$$

Hence, for sufficiently large ν, and $p < q$,

$$|g(x_p, t_o/q_\nu) - g(y_p, t_o/q_\nu)| \leq |x_p - y_p|(1 + Mt_o/q_\nu).$$

Further, by (2),

$$|g(x_p, t_o/q_\nu) - \sigma_\nu(x_p)| < \delta/q_\nu .$$

Since $\sigma_\nu(x_p) = x_{p+1}$ and $g(y_p, t_o/q_\nu) = y_{p+1}$, we obtain

$$|x_{p+1} - y_{p+1}| < \frac{\delta_\nu}{q_\nu} + |x_p - y_p|(1 + M\frac{t_o}{q_\nu}),\ 1 \leq p \leq q_\nu - 1.$$

It follows by iteration that

$$|y_{q_\nu} - x_{q_\nu}| < (\delta_\nu + |x_1 - y_1|)(1 + M\frac{t_o}{q_\nu})^{q_\nu} < 2\delta_\nu e^{Mt_o}$$

(because of (2)). In other words

$$|g(x, t_o) - \sigma_\nu^{q_\nu}(x)| < 2\delta_\nu e^{Mt_o}, \quad \nu \text{ large, } x \in B.$$

It follows that $\sigma_\nu^{q_\nu} \to g_{t_o}$ uniformly on B. As we have already seen, this is sufficient.

<u>Corollary.</u> If X, Y are vector fields associated to G and a, b are real, then so are

$$aX + bY \quad \text{and} \quad [X, Y].$$

Proof. Let $g, h: D \times \mathbb{R} \to D$ be the one-parameter groups associated to X, Y respectively (Corollary to Proposition 4). Then

$$aX(x) + bY(x) = \lim_{k \to \infty} k\{g(h(x, \frac{b}{k}), \frac{a}{k}) - x\}$$

uniformly for x in any compact subset of D. In the same way, if we denote by g_t, h_t respectively the maps $x \to g(x, t)$, $y \to h(x, t)$, and set

$$\sigma_t = g_t \circ h_t \circ g_{-t} \circ h_{-t},$$

then $\lim_{t \to 0} t^{-2}\{\sigma_t(x) - x\} = [X, Y](x)$. We can now apply Theorem 2.

Proposition 5. The (real) vector space of vector fields assoicated to G is finite-dimensional of dimension $\leq 2n(n+1)$.

Proof. Let $X^1, \ldots, X^m$ be linearly independent vector fields associated to G. By Lemma e in §1, there is an analytic map

$$F: \Omega_o \times U \to D$$

($\Omega_o \subset\subset D$, U a suitable neighborhood of 0 in $\mathbb{R}^m$) such that

$$\frac{\partial F}{\partial t}(x, ta) = \sum_{j=1}^{m} a_j X^j(F(x, ta)), \quad F(x, 0) = x.$$

Moreover, if U is small enough, the map

$$a \mapsto F_a, \quad F_a(x) = F(x, a)$$

is injective.

Now, for any $a \in \mathbb{R}^m$, $\sum_{j=1}^{m} a_j X^j$ is a vector field associated to G (by the corollary above); also $t \mapsto F_{ta}$ is the local one-parameter group group associated to $\sum a_j X^j$. Hence, if t is small enough, the map $x \mapsto F(x, ta)$ is the restriction to Ω_o of an element $\tau_t \in G$. If $p > 0$ is large enough, then $\sigma_a = (\tau_{1/p})^p \in G$ and $\sigma_a|\Omega_o = F_a$. This gives us an injection $\sigma: U \to G$, $a \mapsto \sigma_a$. Moreover, since $\sigma_a|\Omega_o = F_a$, it follows from Proposition 3 that σ is continuous. By Proposition 2, G is homeomorphic to a closed subset of $D \times GL(n, \mathbb{C})$, which is an open set in $\mathbb{R}^{2n(n+1)}$. It follows from a classical result of dimension theory that $m \leq 2n(n+1)$. [See e.g., Hurewicz-Wallman, Dimension Theory, Princeton University Press.]

Remark. If we look at the map $\psi: G \to D$ given by $\psi(\sigma) = \sigma(x_o)$ for some $x_o \in D$, it is easily seen that $\psi^{-1}(x)$ is a coset of a compact (Lie) subgroup of $GL(n, \mathbb{C})$, so has real dimension $\leq n^2$. This can be used to show that the dimension of the space of vector fields associated to G is $\leq n(n+2)$. This is best possible; the dimension $= n(n+2)$ when D is the unit ball $D = \{z \in \mathbb{C}^n \mid |z_1|^2 + \ldots + |x_n|^2 < 1\}$. Actually, only for domains holomorphically isomorphic to the ball is this bound attained. See Kaup [18]. Kaup's results are in fact much more precise than this.

Let $D_o \subset\subset D_1 \subset\subset D$, D_o and D_1 being open sets and define for $\sigma \in G$,

$$\mu(\sigma) = \sup_{x \in D_1} |\sigma(x) - x|.$$

Let $0 < \alpha < 1 < \beta$.

Proposition 6. There exists $\rho > 0$ such that the following holds: If $\sigma \in G$ and $q \geq 1$ are such that $\mu(\sigma^p) < \rho$ for $p = 1, \ldots, q-1$, then for $x \in D_o$, we have

$$\alpha q |\sigma(x) - x| \leq |\sigma^q(x) - x| \leq \beta q |\sigma(x) - x|.$$

Proof. Let $D_o \subset\subset D_o' \subset\subset \Omega \subset\subset D_1$. We assert that there exist $\varepsilon_o > 0$ and a constant $C > 0$ such that for any holomorphic mapping $f: D_1 \to \mathbb{C}^n$ with

$$\mu(f) = \sup_{x \in D_1} |f(x) - x| < \varepsilon, \qquad 0 < \varepsilon < \varepsilon_o$$

we have

$$(1 - C\varepsilon)|x-y| \leq |f(x) - f(y)| \leq (1 + C\varepsilon)|x-y|$$

for $x, y \in D_o'$.

In fact, let $g(x) = f(x) - x$. By Cauchy's inequalities and the mean value theorem, there is $C > 0$ such that

$$|g(x) - g(y)| \leq C\varepsilon\, |x-y|\, , \quad x, y \in D_o' .$$

Clearly

$$|x-y| - |g(x) - g(y)| \leq |f(x) - f(y)| \leq |x-y| + |g(x) - g(y)|.$$

The assertion follows immediately from this.

If $\mu(\sigma^p) < \rho$, $p = 1, \ldots, q-1$, and if $f(x) = \frac{1}{q}\{x + \sigma(x) + \ldots + \sigma^{q-1}(x)\}$, then $\mu(f) < \rho$. We choose ρ such that $\sigma(D_o) \subset D_o'$, $\alpha < 1 - C\rho$ and $1 + C\rho < \beta$. Proposition 6 follows at once if we apply the above inequality to the pair of pair of points $x, y = \sigma(x)$.

We come now to the second important step in Cartan's proof.

<u>Theorem 3.</u> If $\{\sigma_\nu\}$ is a sequence of elements in G, $\sigma_\nu \neq$ identity, and $\mu(\sigma_\nu) \to 0$, then there is a sequence of integers q_ν, $q_\nu \to \infty$ such that

$$q_\nu(\sigma_\nu(x) - x)$$

has a subsequence which converges uniformly on compact subsets of D to a map $X: D \to \mathbb{C}^n$, $X \neq 0$.

<u>Proof.</u> Let $0 < \alpha < 1 < \beta$ and Ω be any open subset of D, $\Omega \subset\subset D$. By Propositions 3 and 6, there is $\rho > 0$ such that the following holds: If $\sigma \in G$, $N > 0$ and $\mu(\sigma^p) < \rho$, $p = 1, \ldots, N$, then

$$\alpha p|\sigma(x) - x| \leq |\sigma^p(x) - x| \leq \beta p|\sigma(x) - x| \, , \quad p = 1, \ldots, N+1, \; x \in \Omega.$$

Let ρ_Ω be the largest ρ with this property. We claim that if $\sigma \neq$ identity there is $N > 0$ such that $\mu(\sigma^N) \geq \rho_\Omega$; in fact, if

$\mu(\sigma^N) < \rho_\Omega$ for all N, we would have

$$\alpha N |\sigma(x) - x| \leq |\sigma^N(x) - x| \quad , \quad x \in \Omega$$

so that

$$|\sigma(x) - x| \leq \frac{1}{\alpha N} |\sigma^N(x) - x|.$$

Since σ^N is a map of D into itself, so is bounded uniformly in N, this implies that $\sigma(x) = x$ for all $x \in \Omega$, so that σ = identity, a contradiction. In particular, $\rho_\Omega < \infty$ if G is not reduced to the identity.

For $\sigma \in G - \{e\}$, let $q(\sigma, \Omega)$ be the integer N with $\mu(\sigma^p) < \rho_\Omega$, $p = 1, \ldots, N-1$, $\mu(\sigma^N) \geq \rho_\Omega$. We claim that if Ω, Ω' are relatively compact open subsets of D, then there is $c = c(\Omega, \Omega') > 0$ such that the following holds:

For any $\sigma \in G - \{e\}$, we have

$$c^{-1} q(\sigma, \Omega) \leq q(\sigma, \Omega') \leq c q(\sigma, \Omega).$$

It is sufficient to prove this when $\Omega \subset \Omega'$. Then $\rho_{\Omega'} \leq \rho_\Omega$ so that $q(\sigma, \Omega') \leq q(\sigma, \Omega)$ (by definition of $q(\sigma, \Omega)$). If our assertion were false, there would exist a sequence $\{\tau_\nu\} \subset G - \{e\}$ such that, if we set $q_\nu = q(\tau_\nu, \Omega)$, $q'_\nu = q(\tau_\nu, \Omega')$, then $q_\nu / q'_\nu \to \infty$ as $\nu \to \infty$. By passing to a subsequence of $\{\tau_\nu\}$, we may suppose that $\tau_\nu^{q'_\nu}$ converges to a map $f: D \to \mathbb{C}^n$ uniformly on compact sets. Since $\mu(\tau_\nu^{q'_\nu}) \geq \rho_{\Omega'}$, we have $\mu(f) \geq \rho_{\Omega'}$ so that $f \neq$ identity. On the other hand, for $x \in \Omega$, we have

$$|\tau_\nu^{q'_\nu}(x) - x| \leq \beta q'_\nu |\tau_\nu(x) - x| \leq (\beta q'_\nu / q_\nu \alpha) \cdot \alpha q_\nu |\tau_\nu(x) - x|$$
$$\leq (\beta q'_\nu)/(\alpha q_\nu) \cdot |\tau_\nu^{q_\nu}(x) - x| \leq \text{const} \cdot q'_\nu / q_\nu$$

since $\tau_\nu^{q_\nu}$ is a map of D into itself and so is bounded. Since $q'_\nu / q_\nu \to 0$ by hypothesis, this would imply that f = identity. This proves our assertion.

We choose now a fixed domain $D_o \subset\subset D$ and set, for $\sigma \in G - \{e\}$, $q(\sigma) = q(\sigma, D_o)$. If $\sigma_\nu \in G - \{e\}$, $\mu(\sigma_\nu) \to 0$, set $q_\nu = q(\sigma_\nu)$. Since $\mu(\sigma_\nu) \to 0$ implies that $\sigma_\nu \to e$ in G (Proposition 3), we have $q_\nu \to \infty$. Moreover,

$$q_\nu |\sigma_\nu(x) - x| \leq \frac{1}{\alpha} |\sigma_\nu^{q_\nu}(x) - x|.$$

In addition, $q_\nu(\sigma_\nu(x) - x)$ is bounded on <u>any</u> $\Omega \subset\subset D$ (since $q_\nu \leq \text{const} \cdot q(\sigma_\nu, \Omega)$, so that

$$q_\nu |\sigma_\nu(x) - x| \leq \text{const.} |\sigma_\nu^{q_\nu}(x) - x|, \quad x \in \Omega \;).$$

Hence there is a subsequence $\{\nu_p\} \subset \{\nu\}$ such that

$$q_{\nu_p}(\sigma_{\nu_p}(x) - x)$$

converges uniformly on compact subsets of D to a map $X: D \to \mathbf{C}^n$. Since $\mu(\sigma_{\nu_p}^{q_{\nu_p}}) \geq \rho_{D_o}$, we may suppose by passing to a subsequence of $\{\nu_p\}$, if necessary, that $\{\sigma_{\nu_p}^{q_{\nu_p}}\}$ converges to a map $g: D \to \mathbf{C}^n$ with $\mu(g) \geq \rho_{D_o}$. Moreover, by passing to the limit in the inequality

$$|\sigma_{\nu_p}^{q_{\nu_p}}(x) - x| \leq \beta q_{\nu_p} |\sigma_{\nu_p}(x) - x|, \quad x \in D_o$$

we obtain

$$|g(x) - x| \leq \beta \cdot |X(x)|, \quad x \in D_o$$

Since $\mu(g) \geq \rho_{D_o}$, this implies that $X \not\equiv 0$.

This proves Theorem 3.

Let V be the set of all holomorphic maps $X: D \to \mathbf{C}^n$ which are vector fields associated to G. Then, by Proposition 5 and the corollary to Theorem 2, V is a (finite-dimensional) Lie algebra of vector fields. Let $\Omega_o \subset\subset D$ and U be a suitable neighborhood of 0 in V. Let

$$g: \Omega_o \times U \to D$$

be the local Lie group associated to V. We have seen (see proof of

Proposition 5) that for any $u \in U$, there is $\sigma_u \in G$ such that

$$\sigma_u(x) = g(x, u) \quad \text{for } x \in \Omega_o .$$

We define a map $\varphi: U \to G$ by $\varphi(u) = \sigma_u$. It follows at once from Proposition 3 that φ is continuous.

Proposition 7. $\varphi(U)$ is a neighborhood of e (= identity) in G.

Proof. Let $D_o \subset\subset D_1 \subset \Omega_o$ and $\mu(\sigma) = \sup_{x \in D_1} |\sigma(x) - x|$ as above. Let K be a compact symmetric neighborhood of 0, $K \subset U$. Suppose the result false. Then, there exists a sequence $\{\sigma_\nu\} \subset G - \{e\}$, $\sigma_\nu \to e$ as $\nu \to \infty$, $\sigma_\nu \notin \varphi(U)$.

Define

$$\varepsilon_\nu = \inf_{a \in K} \mu(\varphi(a) \cdot \sigma_\nu).$$

Since φ is continuous and K compact, there is an $a_\nu \in K$ such that

$$\varepsilon_\nu = \mu(\varphi(a_\nu) \cdot \sigma_\nu).$$

In particular, $\varepsilon_\nu > 0$ since if $\varepsilon_\nu = 0$, we would have $\sigma_\nu = \varphi(a_\nu)^{-1} = \varphi(-a_\nu) \in \varphi(U))$. Further

$$\varepsilon_\nu \le \mu(\varphi(0) \cdot \sigma_\nu) = \mu(\sigma_\nu) \to 0$$

since $\sigma_\nu \to e$.

Set

$$\tau_\nu = \varphi(a_\nu) \cdot \sigma_\nu .$$

Since $\mu(\tau_\nu) = \varepsilon_\nu \to 0$, it follows from Theorem 3 that there is a sequence $\{q_\nu\}$ of integers, $q_\nu \to 0$ such that

$$q_\nu(\tau_\nu(x) - x)$$

has a subsequence converging uniformly on compact subsets of D to a map $X: D \to \mathbb{C}^n$, $X \not\equiv 0$. Replacing $\{\tau_\nu\}$ by the corresponding subsequence, we suppose that $q_\nu(\tau_\nu - e) \to X$. By Theorem 2, X is a vector field associated to G.

Let

$$A = \sup_{x \in D_1} |X(x)| .$$

Then

$$q_\nu \varepsilon_\nu = \sup_{x \in D_1} q_\nu |\tau_\nu(x) - x| \to A > 0 \text{ as } \nu \to \infty .$$

Let U_0 be a neighborhood of 0 in V, $\overline{U}_0 \subset \overset{o}{K}$. Then

$$\delta = \inf_{a \in K - U_0} \mu(\varphi(a)) > 0$$

and

$$\lim_{\nu \to \infty} \inf_{a \in K - U_0} \mu(\varphi(a)\sigma_\nu) = \delta$$

since $\sigma_\nu \to e$. Since $\varepsilon_\nu \to 0$, it follows that for sufficiently large ν, $a_\nu \in U_0$.

Let $t \mapsto h_t$ be the one-parameter group of X (corollary to Proposition 4). Then, for small t, $h_t \in \varphi(K)$. Consider

$$\omega_\nu = h_{-1/q_\nu} \circ \tau_\nu .$$

Then $\omega_\nu = h_{-1/q_\nu} \circ \varphi(a_\nu) \circ \sigma_\nu = \varphi(b_\nu) \circ \sigma_\nu$, where $b_\nu = w(-1/q_\nu, a_\nu)$ (in the notation of Theorem 1). Since $a_\nu \in U_0$ for large ν and $q_\nu \to \infty$, $b_\nu \in K$ if ν is large, hence

$$\mu(\omega_\nu) \geq \varepsilon_\nu \quad \text{for large } \nu.$$

We now assert that if

$$\psi_\nu(x) = h_{-1/q_\nu}(\tau_\nu(x)) - x,$$

then $q_\nu \psi_\nu(x) \to 0$ as $\nu \to \infty$ uniformly for $x \in D_1$. In fact, there is a constant $C > 0$ such that

$$\psi_\nu(x) = \tau_\nu(x) - x - \frac{1}{q_\nu} X(\tau_\nu(x)) + q_\nu^{-2} a_\nu(x)$$

where $|a_\nu(x)| \leq C$ for all ν and $x \in D_1$. This gives

$$q_\nu \psi_\nu(x) = q_\nu(\tau_\nu(x) - x) - X(\tau_\nu(x)) + O(q_\nu^{-1}).$$

Since $q_\nu(\tau_\nu(x) - x) \to X(x)$ and $X(\tau_\nu(x)) \to X(x)$ uniformly on compact sets of D, it follows that $q_\nu \psi_\nu(x) \to 0$ uniformly on D_1. Since $\mu(\omega_\nu) \geq \varepsilon_\nu$, this implies that $q_\nu \varepsilon_\nu \to 0$ as $\nu \to \infty$, contradicting our earlier remark that $q_\nu \varepsilon_\nu \to A = \sup_{x \in D_1} |X(x)| > 0$. This proves Proposition 7.

Theorem 4 (H. Cartan). The group $G = \mathrm{Aut}(D)$ carries the structure of a Lie group such that the map $G \times D \to D$, $(\sigma, x) \mapsto \sigma(x)$ is real analytic.

Proof. Let U be as in Proposition 7. If K is a compact neighborhood of 0 in V, $K \subset U$, then $\varphi|K$ is a homeomorphism onto a neighborhood N of e in G. Let W_o be an open neighborhood of e in G such that $W_o = W_o^{-1}$, $W_o \cdot W_o \subset N$ (here $W_o^{-1} = \{\sigma \in G | \sigma^{-1} \in W_o\}$, $W_o \cdot W_o = \{\sigma \cdot \tau | \sigma, \tau \in W_o\}$). Further, $U_o = \varphi^{-1}(W_o)$ is a neighborhood of 0 in V.

We consider the covering $\{W^\sigma\}_{\sigma \in G}$ of G where $W^\sigma = W_o \cdot \sigma = \{\tau \cdot \sigma | \tau \in W_o\}$. Let $\chi_\sigma : W_o \cdot \sigma \mapsto U_o$ be the homeomorphism $\tau \cdot \sigma \to \varphi^{-1}(\tau)$. U_o, being open in the finite-dimensional vector space V, can be looked upon as an open set in $\mathbb{R}^m$, $m = \dim V$. If $W^\sigma \cap W^\tau \neq \emptyset$, the coordinate transformation $\chi_\tau \circ \chi_\sigma^{-1}$ is given by

$$\chi_\tau \circ \chi_\sigma^{-1}(u) = \varphi^{-1}(\varphi(u) \circ \sigma \circ \tau^{-1}).$$

Now $\sigma \circ \tau^{-1} \in N$, so that, with the notation of Theorem 1, if $\sigma \circ \tau^{-1} = \varphi(v_o)$ we have

$$\chi_\tau \circ \chi_\sigma^{-1}(u) = w(u, v_o)$$

which is a real analytic function of u.

If $f: G \times D \to D$ is the map $f(\sigma, x) = \sigma(x)$, then $(\chi_\sigma^{-1} \times \mathrm{id}) \circ f$ is the map $(u, x) \to \varphi(u)(\sigma(x))$ which is analytic (for fixed σ) by Proposition 4.

This proves Cartan's theorem.

The structure of a Lie group on G is uniquely determined by its structure of a topological group. This is a consequence of general theorems of E. Cartan (see e.g., Chevalley [12], pp. 128-129).

The proofs given above follow very closely those of H. Cartan [7]. Cartan's results are even stronger (they give the uniqueness statement above as a by-product, for instance). These methods introduced by Cartan were used by Bochner-Montgomery [4] to prove that a locally compact group acting effectively on a differentiable manifold (i.e., such that no element except the neutral element acts as the identity) and such that each element of the group acts as a diffeomorphism is, in fact, a Lie group. This result of Bochner-Montgomery is, in its turn, basic in the finer study of transformation groups on complex spaces. See Kaup [18] and the references given there.

A final remark. The hypothesis that D be bounded is essential in Cartan's theorem. W. Kaup [18] has given an example of a domain $D \subset \mathbb{C}^n$, unbounded, such that Aut(D) acts transitively on D (i.e., for any pair of points $x, y \in D$, there is $\sigma \in \text{Aut}(D)$ with $\sigma(x) = y$), but nevertheless, no Lie group acts transitively on D. He has also investigated certain classes of spaces which are not equivalent to bounded domains but for which the statement of Cartan's theorem remains valid.

REFERENCES

1. H. Behnke and P. Thullen. Theorie der Funktionen mehreren komplexen Veränderlichen. Erg. der Math. 3. Springer, Berlin, 1934.

2. E. Bishop. Holomorphic completion, analytic continuation, and the interpolation of semi-norms. Annals of Math. 78(1963): 468-500.

3. S. Bochner. A theorem on analytic continuation of functions in several variables. Annals of Math. 39(1938): 14-19.

4. S. Bochner and D. Montgomery. Locally compact groups of differentiable transformations. Annals of Math. 47(1946): 639-53.

5. H. Cartan. Les fontions de deux variables complexes et le problème de la représentation analytique. J. de Math. pures et app. 10(1931): 1-114.

6. H. Cartan. Sur les fonctions de plusieurs variables complexes: L'itération des transformations intérieures d'un domaine borné. Math. Zeit. 35(1932): 760-73.

7. H. Cartan. Sur les groupes de transformations analytiques. Actualités Sc. et Indus. Hermann, Paris, 1935.

8. H. Cartan. Sur les fonctions de n variables complexes: Les transformations du produit topologique de deux domaines bornés. Bull. Soc. math. France 64(1936): 37-48.

9. H. Cartan. Sur une extension d'un théorème de Radó. Math. Annalen 125(1952): 49-50.

10. H. Cartan. Séminaire E.N.S. 1951/52.

11. H. Cartan. Séminaire E.N.S. 1953/54

12. H. Cartan. Séminaire E.N.S. 1960/61

13. H. Cartan and P. Thullen. Zur Theorie der Singularitäten der Funktionen mehrerer komplexer Veränderlichen: Regularitäts- und Konvergenzbereiche. Math. Annalen 106(1932): 617-47.

14. C. Chevalley. Theory of Lie Groups. Princeton Univ. Press. 1946.

15. J. Frenkel. Séminaire sur les théorèmes A et B pour les espaces de Stein. Strasbourg 1965.

16. R. Gunning and H. Rossi. Analytic functions of several complex variables, Prentice Hall, 1965.

17. F. Hartogs. Zur Theorie der analytischen Funktionen mehrerer unabhängiger Veränderlichen insbesondere über die Darstellung derselben durch Reihen, welche nach Potenzen einer Veränderlichen fortschreiten. Math. Annalen 63(1906): 1-88.

18. F. Hartogs. Über die aus der singulären Stellen einer analytischen Funktion mehrerer Veränderlichen bestehenden Gebilde. Acta Math. 32 (1909): 57-79.

19. E. Heinz. Ein elementarer Beweis des Satzes von Radó-Behnke-Stein-Cartan über analytische Funktionen. Math. Annalen 131(1956): 258-59.

20. M. Hervé. Several complex variables: Local theory. Oxford Univ. Press and Tata Institute of Fundamental Research, 1963.

21. L. Hörmander. An introduction to complex analysis in several variables. Van Nostrand, 1966.

22. W. Kaup. Reelle Transformationsgruppen und invariante Metriken auf komplexen Räumen. Inventiones Math. 3(1967): 43-70.

23. B. Malgrange. Lectures on functions of several complex variables. Tata Institute of Fundamental Research, 1958.

24. R. Narasimhan. Introduction to the theory of analytic spaces. Springer Lecture Notes, no. 25, 1966.

25. R. Narasimhan. Analysis on real and complex manifolds. North-Holland, 1968.

26. K. Oka. Sur les fonctions analytiques de plusieurs variables. IX, Domaines finis sans points critiques intérieurs. Jap. J. of Math.

27. F. Osgood. Lehrbuch der Funktionentheorie, vol. 2, pt. 1. B. G. Teubner, Leipzig, 1924.

28. T. Radó. Subharmonic functions. Chelsea Pub. Co., 1949.

29. T. Radó. Über eine nicht-fortsetzbare Riemannsche Mannigfaltigkeit. Math. Zeit. 20(1924): 1-6.

30. K. Reinhardt. Über Abbildungen durch analytische Funktionen Zweier Veränderlichen. Math. Annalen 83(1921): 211-55.

31. R. Remmert and K. Stein. Eigentliche holomorphe Abbildungen. Math. Zeit. 73(1960): 159-89.

32. P. Thullen. Zur Theorie der Singularitäten der Funktionen mehrerer komplexer Veränderlichen: Die Regularitäts-hullen, Math. Annalen 106(1932): 64-76.